© 2018
Clement Ampadu
drampadu@hotmail.com

ISBN: 978-1-387-50300-1
ID: 22401407
www.lulu.com

All rights reserved. No part of this publication may be produced or transmitted in any form or by any means, electronic or mechanical, including photocopying and recording, or in any information storage and retrieval system, without the prior written permission of the author

Contents

Preface   3

Dedication   4

I   Metric and Related Spaces   5

A — Some Results in Metric Space   6

1   Higher-Order Hardy and Rogers Type Common Fixed Point Theorems of r-Compatible Mappings of Type (A) in Multiplicative Metric Space   7
- 1.1   Brief Summary 7
- 1.2   Preliminaries 7
- 1.3   Main Results 8
- 1.4   Open Problems 11

2   Higher-Order Jungck-Type Contraction Mapping Theorem under Faintly r-Compatible Mappings in Metric Space   16
- 2.1   Brief Summary 16
- 2.2   Preliminaries 16
- 2.3   Main Results 20
- 2.4   Open Problems 21

3   A Higher-Order Common Fixed Point Theorem under r-Compatible Mappings of Type (R) in Metric Space   25
- 3.1   Brief Summary 25
- 3.2   Preliminaries 25
- 3.3   Main Results 26
- 3.4   Open Problem 30

4   A Common Higher-Order Fixed Point Theorem under r-Compatible Mappings of Type (P) in Metric Space   31
- 4.1   Brief Summary 31
- 4.2   Preliminaries 31
- 4.3   Main Results 33
- 4.4   Open Problem 36

5   A Higher-Order Fixed Point Theorem under r-Compatible Mappings of Type (K)   37
- 5.1   Brief Summary 37
- 5.2   Preliminaries 37
- 5.3   Main Results 38
- 5.4   Open Problems 43

Bibliography   43

Preface

The commuting mapping concept is popular in fixed point theory, and a generalization of it, called compatible mappings appeared in [GERALD JUNGCK, COMPATIBLE MAPPINGS AND COMMON FIXED POINTS, Internat. J. Math. and Math. Sci. Vol. 9 No. 4 (1986) 771-779]. Since the concept of compatible mappings appeared in the literature, new generalizations of it have been found, and the fixed point theory relating these concepts have been investigated. In this book, inspired by higher-order fixed point theory [Clement Ampadu, Fixed Point Theory for Higher-Order Mappings. ISBN: 5800118959925, lulu.com, 2016] we consider when the r-times composition of two mappings commute, and introduce a new concept called r-compatible mappings, we then investigate some higher-order fixed point theory relating r-compatible mappings and related concepts in the setting of metric and multiplicative metric space.

Inspired by the work contained in [G. Jungck, P.P. Murthy and Y.J. Cho, Compatible mappings of type (A) and common fixed points, Math. Japonica, 38(1993), 381-390], we introduce r-compatible mappings of type (A) in Chapter 1, and investigate the common higher-order fixed point theory for maps satisfying the Hardy and Rogers contraction under r-compatibility of type (A) in the setting of multiplicative metric space.

Chapter 2 is inspired by the work contained in [Ravindra K Bisht and Naseer Shahzad, Faintly compatible mappings and common fixed points, Fixed Point Theory and Applications 2013, 2013:156], and in this chapter we introduce a concept of faintly r-compatible mappings, and investigate the common higher-order fixed point theory for maps satisfying the Banach contraction under faintly r-compatibility in the setting of metric space.

Inspired by the work contained in [Y. Rohen, M.R. Singh and L. Shambhu, Common fixed points of compatible mapping of type (C) in Banach Spaces, Proc. of Math. Soc., BHU 20(2004), 77-87], we introduce r-compatible mappings of type (R) in Chapter 3, and investigate the common higher-order fixed point theory for maps satisfying the contractive condition contained in [M. Koireng Meitei, Leenthoi Ningombam and Yumnam Rohen, Common Fixed Points of Compatible Mappings of Type (R), Gen. Math. Notes, Vol. 10, No. 1, May 2012, pp. 58-62] under r-compatibility of type (R) in the setting of metric space.

Chapter 4 is inspired by the work contained in [V. Srinivas and V. Naga Raju, Common Fixed Point Theorem on Compatible Mappings of Type (P), Gen. Math. Notes, Vol. 21, No. 2, April 2014, pp. 87-94], and in this chapter we introduce r-compatible mappings of type (P), and investigate the common higher-order fixed point theory for maps satisfying the contractive condition contained in [M. Koireng Meitei, Leenthoi Ningombam and Yumnam Rohen, Common Fixed Points of Compatible Mappings of Type (R), Gen. Math. Notes, Vol. 10, No. 1, May 2012, pp. 58-62] under r-compatibility of type (P) in the setting of metric space.

Inspired by the work contained in [Ravi Sriramula and V.Srinivas, A Result on Fixed Point Theorem Using Compatible Mappings of Type (K), Annals of Pure and Applied Mathematics Vol. 13, No. 1, 2017, 41-47], we introduce r-compatible mappings of type (K) in Chapter 5, and investigate the common higher-order fixed point theory for maps satisfying the contractive condition contained in [M. Koireng Meitei, Leenthoi Ningombam and Yumnam Rohen, Common Fixed Points of Compatible Mappings of Type (R), Gen. Math. Notes, Vol. 10, No. 1, May 2012, pp. 58-62] under r-compatibility of type (K) in the setting of metric space.

Dedication

This work is dedicated to the nuclear and extended family, friends, colleagues and other people who have shown interest in loving me.

Clement Boateng Ampadu
January 2018

Part I
Metric and Related Spaces

Subpart A

Some Results in Metric Space

Chapter 1

Higher-Order Hardy and Rogers Type Common Fixed Point Theorems of r-Compatible Mappings of Type (A) in Multiplicative Metric Space

1.1 Brief Summary

> **Abstract**
>
> Inspired by higher-order fixed point theory [Clement Ampadu, Fixed Point Theory for Higher-Order Mappings. ISBN: 5800118959925, lulu.com, 2016] we obtain the higher-order version of Theorem 3.1[M. Bina Devi, Some Common Fixed Point Theorems of Compatible Mappings of Type (A) in Metric Space, Gen. Math. Notes, Vol. 14, No. 1, January 2013, pp. 43-50], in the setting of multiplicative metric space.

1.2 Preliminaries

Taking inspiration from [G. Jungck, Compatible mappings and common fixed points, Internat. J. Math. Math. Sci., 9(1986), 771-779] we introduce the following

> **Definition 1.2.1**
>
> Let S and T be mappings from a multiplicative metric space X into itself. The mappings S and T will be called r-compatible mappings if $\lim_{n\to\infty} m(S^r T^r x_n, T^r S^r x_n) = 1$ whenever $\{x_n\}$ is a sequence in X such that $\lim_{n\to\infty} S^r x_n = \lim_{n\to\infty} T^r x_n = z$ for some $z \in X$

Taking inspiration from [G. Jungck, P.P. Murthy and Y.J. Cho, Compatible mappings of type (A) and common fixed points, Math. Japonica, 38(1993), 381-390] we introduce the following

> **Definition 1.2.2**
>
> Let S and T be mappings from a multiplcative metric space X into itself. The mappings S and T will be called r-compatible mappings of type (A) if $\lim_{n\to\infty} m(S^r T^r x_n, T^r T^r x_n) = 1$ and $\lim_{n\to\infty} m(T^r S^r x_n, S^r S^r x_n) = 1$ whenever $\{x_n\}$ is a sequence in X such that $\lim_{n\to\infty} S^r x_n = \lim_{n\to\infty} T^r x_n = z$ for some $z \in X$

In the sequel we will need the following Propositions whose metric counterpart, when $r = 1$, can be found in [Y. Rohen, Th. Indubala, O. Budhichandra and N. Leenthoi, Common fixed point theorems for compatible mappings of type (A), IJMSEA, 6(II) (2012), 323-333]

Proposition 1.2.3

Let S and T be r-continuous mappings from a multiplicative metric space X into itself, that is, S^r and T^r are continuous for any $r \in \mathbb{N}$. S and T are r-compatible iff they are r-compatible mappings of type (A)

Proposition 1.2.4

Let S and T be mappings from a multiplicative metric space (X, m) into itself. If a pair $\{S, T\}$ is r-compatible of type (A) on X and $\lim_{n \to \infty} S^r x_n = \lim_{n \to \infty} T^r x_n = z$ for some $z \in X$, then we have the following

(a) $\lim_{n \to \infty} m(T^r S^r x_n, S^r z) = 1$, if S is r-continuous, that is, S^r is continuous for any $r \in \mathbb{N}$

(b) $\lim_{n \to \infty} m(S^r T^r x_n, T^r z) = 1$, if T is r-continuous, that is, T^r is continuous for any $r \in \mathbb{N}$

(c) $S^r T^r z = T^r S^r z$ and $S^r z = T^r z$, if S and T are r-continuous at z, that is, S^r and T^r are continuous at z for any $r \in \mathbb{N}$.

1.3 Main Results

Definition 1.3.1

Let S, T, and A be three self-maps of a metric space (X, d) into itself. We say the pair (S, T) is Hardy and Rogers type contractive with respect to A if the following holds for all $x, y \in X$ and $0 \leq k < \frac{1}{6}$

$$d(Sx, Ty) \leq k[d(Ax, Ay) + d(Sx, Ax) + d(Ty, Ay) + d(Ax, Ty) + d(Ay, Sx)]$$

Definition 1.3.2

Let S, T, and A be three self-maps of a metric space (X, d) into itself. We say the pair (S, T) is a higher-order Hardy and Rogers type contraction with respect to A if the following holds for all $x, y \in X$

$$d(S^r x, T^r y) \leq \sum_{q=0}^{r-1} c_q [d(A^{q+1} x, A^{q+1} y) + d(S^{q+1} x, A^{q+1} x) + d(T^{q+1} y, A^{q+1} y)$$
$$+ d(A^{q+1} x, T^{q+1} y) + d(A^{q+1} y, S^{q+1} x)]$$

where $0 \leq c_q < \frac{1}{6}$, for all $0 \leq q \leq r-1$, and $r \in \mathbb{N}$

Proposition 1.3.3

Let (X, d) be metric space, and $S, T, A : X \mapsto X$, such that the pair (S, T) is a higher-order Hardy and Rogers type contraction with respect to A. For every pair $x \neq y$, define

$$Z := Z(x, y) = \max_{0 \leq v \leq r-1} \beta^{-v} \frac{d(S^v x, T^v y)}{d(Ax, Ay) + d(Sx, Ax) + d(Ty, Ay) + d(Ax, Ty) + d(Ay, Sx)}$$

then

$$Z = \max_{n \in \mathbb{N} \cup \{0\}} \beta^{-n} \frac{d(S^n x, T^n y)}{d(Ax, Ay) + d(Sx, Ax) + d(Ty, Ay) + d(Ax, Ty) + d(Ay, Sx)}$$

where $\beta \in [0, \frac{1}{6})$

Now using the above Proposition, we have the following alternate characterization of Definition 1.3.2

> **Definition 1.3.4**
>
> Let S, T, and A be three self-maps of a metric space (X,d) into itself. We say the pair (S,T) is higher-order Hardy and Rogers type contraction with respect to A if the following holds for all $x, y \in X$ and $r \in \mathbb{N}$
>
> $$d(S^r x, T^r y) \leq Z\beta^r [d(Ax, Ay) + d(Sx, Ax) + d(Ty, Ay) + d(Ax, Ty) + d(Ay, Sx)]$$
>
> where $Z \geq 1$ is given by the previous Proposition, and $\beta \in [0, \frac{1}{6})$.

Now the multiplicative version of the above reads as follows

> **Definition 1.3.5**
>
> Let S, T, and A be three self-maps of a multiplicative metric space (X,m) into itself. We say the pair (S,T) is higher-order Hardy and Rogers type multiplicative contraction with respect to A if the following holds for all $x, y \in X$ and $r \in \mathbb{N}$
>
> $$m(S^r x, T^r y) \leq [m(Ax, Ay) \cdot m(Sx, Ax) \cdot m(Ty, Ay) \cdot m(Ax, Ty) \cdot m(Ay, Sx)]^{Z\beta^r}$$
>
> where $Z \geq 1$ is given by the previous Proposition, and $\beta \in [0, \frac{1}{6})$.

Now our main result is the following

> **Theorem 1.3.6**
>
> Suppose S, T, A are three self mappings of a multiplicative complete multiplicative metric space (X, m) into itself, satisfying the following conditions
>
> (a) $S^r(X) \cup T^r(X) \subset A^r(X)$ for any $r \in \mathbb{N}$
>
> (b) the pair (S,T) is higher-order Hardy and Rogers type multiplicative contraction with respect to A
>
> (c) one of S, T, and A is r-continuous
>
> (d) (S, A) and (T, A) are r-compatible mappings of type (A)
>
> Then S, T, and A have a unique common r-fixed point.

Proof of Theorem 1.3.6

Let $x_0 \in X$ be arbitrary and construct a sequence $\{x_n\}$ in X as follows, for all $r \in \mathbb{N}$ and $n = 0, 1, 2, \cdots$

$$A^r x_{2n+1} = S^r x_{2n}, \quad A^r x_{2n+2} = T^r x_{2n+1}$$

Now observe we have the following

$$\begin{aligned} m(A^r x_{2n+1}, A^r x_{2n+2}) &= m(S^r x_{2n}, T^r x_{2n+1}) \\ &\leq [m(A^r x_{2n+1}, A^r x_{2n+2}) \cdot m(S^r x_{2n}, A^r x_{2n}) \cdot m(T^r x_{2n+1}, A^r x_{2n+1}) \\ &\quad \cdot m(A^r x_{2n}, T^r x_{2n+1}) \\ &\quad \cdot m(A^r x_{2n+1}, S^r x_{2n})]^{Z\beta^r} \\ &= [m(A^r x_{2n}, A^r x_{2n+1})^2 \cdot m(A^r x_{2n+2}, A^r x_{2n+1}) \cdot m(A^r x_{2n}, A^r x_{2n+2}) \\ &\quad \cdot m(A^r x_{2n+1}, A^r x_{2n+1})]^{Z\beta^r} \\ &\leq m(A^r x_{2n}, A^r x_{2n+2})^{3Z\beta^r} \cdot m(A^r x_{2n+1}, A^r x_{2n+2})^{2Z\beta^r} \end{aligned}$$

From the above, we deduce that

$$m(A^r x_{2n+1}, A^r x_{2n+2}) \leq m(A^r x_{2n}, A^r x_{2n+1})^{\frac{3Z\beta^r}{1-2Z\beta^r}}$$

Now put $h := \frac{3Z\beta^r}{1-2Z\beta^r} < 1$, then from the inequality immediately above, and by mathematical induction, we deduce that

$$m(A^r x_{2n+1}, A^r x_{2n+2}) \leq m(A^r x_0, A^r x_1)^{h^{2n+1}}$$

Since $h < 1$, consequently, the sequence $\{A^r x_n\}$ is multiplicative Cauchy. Since X is multiplicative complete, there exists $z \in X$ such that $\lim_{n \to \infty} A^r x_n = z$. Since $\{T^r x_{2n+1}\}$ and $\{S^r x_{2n}\}$ are subsequences of $\{A^r x_{2n}\}$. It follows that $\lim_{n \to \infty} S^r x_{2n} = z$ and $\lim_{n \to \infty} T^r x_{2n+1} = z$. Suppose that A is r-continuous and the pair (S, A) is r-compatible of type (A), then, observe we have the following

$$\begin{aligned} m(S^r A^r x_{2n}, T^r x_{2n+1}) &\leq [m(A^r A^r x_{2n}, A^r x_{2n+1}) \cdot m(S^r A^r x_{2n}, A^r A^r x_{2n}) \\ &\quad \cdot m(T^r x_{2n+1}, A^r x_{2n+1}) \cdot m(A^r A^r x_{2n}, T^r x_{2n+1}) \\ &\quad \cdot m(A^r x_{2n+1}, S^r A^r x_{2n})]^{Z\beta^r} \end{aligned}$$

Since A is r-continuous and the pair (S, A) is r-compatible of type (A), then taking limits in the above as $n \to \infty$ and using the facts that $\lim_{n \to \infty} A^r A^r x_{2n} = A^r z$, and $\lim_{n \to \infty} S^r A^r x_{2n} = A^r z$, we deduce the following

$$m(A^r z, z) \leq m(A^r z, z)^{3Z\beta^r}$$

However $1 - 3Z\beta^r \neq 0$, thus the above implies that $m(A^r z, z) = 1$, that is, $A^r z = z$. Now observe we have the following

$$\begin{aligned} m(S^r z, T^r x_{2n+1}) &\leq [m(A^r A^r z, A^r x_{2n+1}) \cdot m(S^r z, A^r A^r z) \cdot m(T^r x_{2n+1}, A^r x_{2n+1}) \\ &\quad \cdot m(A^r z, T^r x_{2n+1}) \cdot m(A^r x_{2n+1}, S^r z)]^{Z\beta^r} \end{aligned}$$

Now taking limits in the above as $n \to \infty$ and using the fact that $A^r z = z$, we deduce the following

$$m(S^r z, z) \leq m(S^r, z)^{2Z\beta^r}$$

However, $1 - 2Z\beta^r \neq 0$, thus the above implies $S^r z = z$, and hence $A^r z = S^r z = z$. Now observe we have the following

$$\begin{aligned} m(z, T^r z) &= m(S^r z, T^r z) \\ &\leq [m(A^r z, A^r z) \cdot m(S^r z, A^r z)^2 \cdot m(T^r z, A^r z)^2 \cdot]^{Z\beta^r} \\ &= m(T^r z, z)^{2Z\beta^r} \end{aligned}$$

> **Proof of Theorem 1.3.6 Continued**
>
> From the above and since $1 - 2Z\beta^r \neq 0$, we deduce that $m(z, T^r z) = 1$, that is, $z = T^r z$. It follows that z is a common r-fixed point of S, T, and A. In a similar fashion one can show that z is a common r-fixed point of S, T, and A when the pair (T, A) is r-compatible of type (A). Finally we show uniqueness of the r-fixed point. Suppose $a = S^r a = T^r a = A^r a$, $b = S^r b = T^r b = A^r b$, but $a \neq b$, then observe we have the following
>
> $$\begin{aligned} m(a,b) &= m(S^r a, T^r b) \\ &\leq [m(A^r a, T^r b) \cdot m(S^r a, A^r a) \cdot m(T^r b, A^r b) \cdot m(A^r a, T^r b) \cdot m(A^r b, S^r a)]^{Z\beta^r} \\ &= [m(a,b)^3 \cdot m(a,a) \cdot m(b,b)]^{Z\beta^r} \\ &= m(a,b)^{3Z\beta^r} \end{aligned}$$
>
> Since $1 - 3Z\beta^r \neq 0$, it follows from the inequality immediately above that $m(a,b) = 1$, that is, $a = b$, and uniqueness follows, and the proof is finished

1.4 Open Problems

Problem 1.4.1

> **Definition 1.4.1.1**
>
> Let S, T, A, and B be four self-maps of a metric space (X, d) into itself. We say the pair (S, T) is Hardy and Rogers type contractive with respect to (A, B) if the following holds for all $x, y \in X$ and $0 \leq k < \frac{1}{6}$
>
> $$d(Sx, Ty) \leq k[d(Ax, By) + d(Ax, Sx) + d(By, Ty) + d(Ax, Ty) + d(By, Sx)]$$

> **Definition 1.4.1.2**
>
> Let S, T, A and B be four self-maps of a metric space (X, d) into itself. We say the pair (S, T) is a higher-order Hardy and Rogers type contraction with respect to (A, B) if the following holds for all $x, y \in X$
>
> $$\begin{aligned} d(S^r x, T^r y) \leq \sum_{q=0}^{r-1} c_q [&d(A^{q+1}x, B^{q+1}y) + d(A^{q+1}x, S^{q+1}x) + d(B^{q+1}y, T^{q+1}y) \\ &+ d(A^{q+1}x, T^{q+1}y) + d(B^{q+1}y, S^{q+1}x)] \end{aligned}$$
>
> where $0 \leq c_q < \frac{1}{6}$, for all $0 \leq q \leq r-1$, and $r \in \mathbb{N}$

> **Proposition 1.4.1.3**
>
> Let (X, d) be metric space, and $S, T, A, B : X \mapsto X$, such that the pair (S, T) is a higher-order Hardy and Rogers type contraction with respect to (A, B). For every pair $x \neq y$, define
>
> $$Z := Z(x,y) = \max_{0 \leq v \leq r-1} \beta^{-v} \frac{d(S^v x, T^v y)}{d(Ax, By) + d(Ax, Sx) + d(By, Ty) + d(Ax, Ty) + d(By, Sx)}$$
>
> then
>
> $$Z = \max_{n \in \mathbb{N} \cup \{0\}} \beta^{-n} \frac{d(S^n x, T^n y)}{d(Ax, By) + d(Ax, Sx) + d(By, Ty) + d(Ax, Ty) + d(By, Sx)}$$
>
> where $\beta \in [0, \frac{1}{6})$

Now using the above Proposition, we have the following alternate characterization of Definition 1.4.1.2

> **Definition 1.4.1.4**
>
> Let S, T, A and B be four self-maps of a metric space (X, d) into itself. We say the pair (S, T) is higher-order Hardy and Rogers type contraction with respect to (A, B) if the following holds for all $x, y \in X$ and $r \in \mathbb{N}$
>
> $$d(S^r x, T^r y) \leq Z\beta^r [d(Ax, By) + d(Ax, Sx) + d(By, Ty) + d(Ax, Ty) + d(By, Sx)]$$
>
> where $Z \geq 1$ is given by the previous Proposition, and $\beta \in [0, \frac{1}{6})$

Now the multiplicative version of the above reads as follows

> **Definition 1.4.1.5**
>
> Let S, T, A, and B be four self-maps of a multiplicative metric space (X, m) into itself. We say the pair (S, T) is higher-order Hardy and Rogers type multiplicative contraction with respect to (A, B) if the following holds for all $x, y \in X$ and $r \in \mathbb{N}$
>
> $$m(S^r x, T^r y) \leq [m(Ax, By) \cdot m(Ax, Sx) \cdot m(By, Ty) \cdot m(Ax, Ty) \cdot m(By, Sx)]^{Z\beta^r}$$
>
> where $Z \geq 1$ is given by the previous Proposition, and $\beta \in [0, \frac{1}{6})$

The open problem is to prove the following

> **Theorem 1.4.1.6**
>
> Let S, T, A, B be four self mappings of a multiplicative complete multiplicative metric space (X, m) into itself, satisfying the following conditions
>
> (a) $S^r(X) \cup T^r(X) \subset A^r(X) \cup B^r(X)$ for any $r \in \mathbb{N}$
>
> (b) the pair (S, T) is higher-order Hardy and Rogers type multiplicative contraction with respect to (A, B)
>
> (c) one of S, T, A, and B is r-continuous
>
> (d) (S, A) and (T, B) are r-compatible mappings of type (A)
>
> Then S, T, A and B have a unique common r-fixed point.

Problem 1.4.2

> **Definition 1.4.2.1**
>
> Let S, T, and A be three self-maps of a metric space (X, d) into itself. We say the pair (S, T) is a Banach contraction with respect to A if the following holds for all $x, y \in X$ and $0 \leq k < 1$
>
> $$d(Sx, Ty) \leq k d(Ax, Ay)$$

> **Definition 1.4.2.2**
>
> Let S, T, and A be three self-maps of a metric space (X, d) into itself. We say the pair (S, T) is a higher-order Banach contraction with respect to A if the following holds for all $x, y \in X$
>
> $$d(S^r x, T^r y) \leq \sum_{q=0}^{r-1} c_q d(A^{q+1} x, A^{q+1} y)$$
>
> where $0 \leq c_q < 1$, for all $0 \leq q \leq r-1$, and $r \in \mathbb{N}$

> **Proposition 1.4.2.3**
>
> Let (X, d) be metric space, and $S, T, A : X \mapsto X$, such that the pair (S, T) is a higher-order Banach contraction with respect to A. For every pair $x \neq y$, define
>
> $$Z := Z(x, y) = \max_{0 \leq v \leq r-1} \beta^{-v} \frac{d(S^v x, T^v y)}{d(Ax, Ay)}$$
>
> then
>
> $$Z = \max_{n \in \mathbb{N} \cup \{0\}} \beta^{-n} \frac{d(S^n x, T^n y)}{d(Ax, Ay)}$$
>
> where $\beta \in [0, 1)$

Now using the above Proposition, we have the following alternate characterization of Definition 1.4.2.2

> **Definition 1.4.2.4**
>
> Let S, T, and A be three self-maps of a metric space (X, d) into itself. We say the pair (S, T) is higher-order Banach contraction with respect to A if the following holds for all $x, y \in X$ and $r \in \mathbb{N}$
>
> $$d(S^r x, T^r y) \leq Z \beta^r d(Ax, Ay)$$
>
> where $Z \geq 1$ is given by the previous Proposition, and $\beta \in [0, 1)$

Now the multiplicative version of the above reads as follows

> **Definition 1.4.2.5**
>
> Let S, T, and A be three self-maps of a multiplicative metric space (X, m) into itself. We say the pair (S, T) is higher-order Banach multiplicative contraction with respect to A if the following holds for all $x, y \in X$ and $r \in \mathbb{N}$
>
> $$m(S^r x, T^r y) \leq m(Ax, Ay)^{Z \beta^r}$$
>
> where $Z \geq 1$ is given by the previous Proposition, and $\beta \in [0, 1)$

Now the open problem is the following

> **Theorem 1.4.2.6**
>
> Suppose S, T, A are three self mappings of a multiplicative complete multiplicative metric space (X, m) into itself, satisfying the following conditions
>
> (a) $S^r(X) \cup T^r(X) \subset A^r(X)$ for any $r \in \mathbb{N}$
>
> (b) the pair (S, T) is a higher-order Banach multiplicative contraction with respect to A
>
> (c) one of S, T, and A is r-continuous
>
> (d) (S, A) and (T, A) are r-compatible mappings of type (A)
>
> Then S, T, and A have a unique common r-fixed point.

Problem 1.4.3

> **Definition 1.4.3.1**
>
> Let S, T, A, and B be four self-maps of a metric space (X, d) into itself. We say the pair (S, T) is a Banach contraction with respect to (A, B) if the following holds for all $x, y \in X$ and $0 \leq k < 1$
>
> $$d(Sx, Ty) \leq k d(Ax, By)$$

Definition 1.4.3.2

Let S, T, A and B be four self-maps of a metric space (X, d) into itself. We say the pair (S, T) is a higher-order Banach contraction with respect to (A, B) if the following holds for all $x, y \in X$

$$d(S^r x, T^r y) \leq \sum_{q=0}^{r-1} c_q d(A^{q+1} x, B^{q+1} y)$$

where $0 \leq c_q < 1$, for all $0 \leq q \leq r-1$, and $r \in \mathbb{N}$

Proposition 1.4.3.3

Let (X, d) be metric space, and $S, T, A, B : X \mapsto X$, such that the pair (S, T) is a higher-order Banach contraction with respect to (A, B). For every pair $x \neq y$, define

$$Z := Z(x, y) = \max_{0 \leq v \leq r-1} \beta^{-v} \frac{d(S^v x, T^v y)}{d(Ax, By)}$$

then

$$Z = \max_{n \in \mathbb{N} \cup \{0\}} \beta^{-n} \frac{d(S^n x, T^n y)}{d(Ax, By)}$$

where $\beta \in [0, 1)$

Now using the above Proposition, we have the following alternate characterization of Definition 1.4.3.2

Definition 1.4.3.4

Let S, T, A and B be four self-maps of a metric space (X, d) into itself. We say the pair (S, T) is higher-order Banach contraction with respect to (A, B) if the following holds for all $x, y \in X$ and $r \in \mathbb{N}$

$$d(S^r x, T^r y) \leq Z \beta^r d(Ax, By)$$

where $Z \geq 1$ is given by the previous Proposition, and $\beta \in [0, 1)$

Now the multiplicative version of the above reads as follows

Definition 1.4.3.5

Let S, T, A, and B be four self-maps of a multiplicative metric space (X, m) into itself. We say the pair (S, T) is a higher-order Banach multiplicative contraction with respect to (A, B) if the following holds for all $x, y \in X$ and $r \in \mathbb{N}$

$$m(S^r x, T^r y) \leq m(Ax, By)^{Z \beta^r}$$

where $Z \geq 1$ is given by the previous Proposition, and $\beta \in [0, 1)$

The open problem is to prove the following

> **Theorem 1.4.3.6**
>
> Let S, T, A, B be four self mappings of a multiplicative complete multiplicative metric space (X, m) into itself, satisfying the following conditions
>
> (a) $S^r(X) \cup T^r(X) \subset A^r(X) \cup B^r(X)$ for any $r \in \mathbb{N}$
>
> (b) the pair (S, T) is a higher-order Banach multiplicative contraction with respect to (A, B)
>
> (c) one of S, T, A, and B is r-continuous
>
> (d) (S, A) and (T, B) are r-compatible mappings of type (A)
>
> Then S, T, A and B have a unique common r-fixed point.

Chapter 2

Higher-Order Jungck-Type Contraction Mapping Theorem under Faintly r-Compatible Mappings in Metric Space

2.1 Brief Summary

Abstract

Inspired by higher-order fixed point theory [Clement Ampadu, Fixed Point Theory for Higher-Order Mappings. ISBN: 5800118959925, lulu.com, 2016] and the notion of faintly compatible mappings [Ravindra K Bisht and Naseer Shahzad, Faintly compatible mappings and common fixed points, Fixed Point Theory and Applications 2013, 2013:156], we introduce faintly r-compatible mappings and obtain the higher-order version of some common fixed point theorems inspired by the Jungck fixed point theorem [G. Jungck, "Commuting mappings and fixed points," American Mathematical Monthly, vol. 83, no. 4, pp. 261–263, 1976]

2.2 Preliminaries

Inspired by compatible mappings [Jungck, G: Compatible mappings and common fixed points. Int. J. Math. Math. Sci. 9(4), 771-779 (1986)] we introduce the following

Definition 2.2.1

A pair of self-maps (A, S) of a metric space (X, d) will be called r-compatible iff

$$\lim_{n \to \infty} d(A^r S^r y_n, S^r A^r y_n) = 0$$

whenever $\{y_n\}$ is a sequence in X such that $\lim_{n \to \infty} A^r y_n = \lim_{n \to \infty} S^r y_n = t$ for some $t \in X$ and any $r \in \mathbb{N}$

Inspired by noncompatible mappings [Ravindra K Bisht and Naseer Shahzad, Faintly compatible mappings and common fixed points, Fixed Point Theory and Applications 2013, 2013:156] we introduce the following

> **Definition 2.2.2**
>
> A pair of self-maps (A, S) of a metric space (X, d) will be called r-noncompatible if there exists a sequence $\{y_n\}$ in X such that $\lim_{n\to\infty} A^r y_n = \lim_{n\to\infty} S^r y_n = t$ for some $t \in X$ and any $r \in \mathbb{N}$, but $\lim_{n\to\infty} d(A^r S^r y_n, S^r A^r y_n) \neq 0$ or $\lim_{n\to\infty} d(A^r S^r y_n, S^r A^r y_n)$ does not exist for any $r \in \mathbb{N}$.

Inspired by weakly compatible maps [Jungck, G: Common fixed points for noncontinuous nonself maps on nonmetric spaces. Far East J. Math. Sci. 4, 199-215 (1996)] we introduce the following

> **Definition 2.2.3**
>
> A pair of self-maps (A, S) of a metric space (X, d) will be called weakly r-compatible, if the pair r-commutes on the set of r-coincidence points, that is, $A^r(S^r(x)) = S^r(A^r(x))$, whenever x is an r-coincidence point of (A, S), that is, $A^r(x) = S^r(x)$ for $x \in X$ and any $r \in \mathbb{N}$

Inspired by occasionally weakly compatible mappings [Al-Thagafi, MA, Shahzad, N: Generalized I-nonexpansive selfmaps and invariant approximations. Acta Math. Sin. 24, 867-876 (2008)] we introduce the following

> **Definition 2.2.4**
>
> A pair of self-maps (A, S) of a metric space (X, d) will be called occasionally weakly r-compatible if there exists an r-coincidence point $x \in X$ such that $A^r(x) = S^r(x)$ implies $A^r(S^r(x)) = S^r(A^r(x))$ for any $r \in \mathbb{N}$

Inspired by conditionally commuting mappings [Pant, V, Pant, RP: Common fixed points of conditionally commuting maps. Fixed Point Theory 1, 113-118 (2010)] we introduce the following

> **Definition 2.2.5**
>
> A pair of self-maps (A, S) of a metric space (X, d) will be called conditionally r-commuting if the pair r-commutes on a nonempty subset of the set of r-coincidence points whenever the set of r-coincidences is nonempty.

Inspired by the concept of subcompatible mappings [Bouhadjera, H, Godet-Thobie, C: Common fixed point theorems for pair of subcompatible maps. arXiv:0906.3159v1 [math. FA] (2009)] we introduce the following

> **Definition 2.2.6**
>
> A pair of self-maps (A, S) of a metric space (X, d) will be called r-subcompatible if there exists a sequence $\{y_n\}$ in X such that $\lim_{n\to\infty} A^r y_n = \lim_{n\to\infty} S^r y_n = t$ for some $t \in X$ and any $r \in \mathbb{N}$, and $\lim_{n\to\infty} d(A^r S^r y_n, S^r A^r y_n) = 0$ for any $r \in \mathbb{N}$

Inspired by conditionally compatible mappings [Pant, RP, Bisht, RK: Occasionally weakly compatible mappings and fixed points. Bull. Belg. Math. Soc. Simon Stevin 19, 655-661 (2012)] we introduce the following

> **Definition 2.2.7**
>
> A pair of self-maps (A, S) of a metric space (X, d) will be called conditionally r-compatible if whenever the set of sequences $\{y_n\}$ satisfying $\lim_{n\to\infty} A^r y_n = \lim_{n\to\infty} S^r y_n$ is nonempty, for any $r \in \mathbb{N}$, there exists a sequence $\{z_n\}$ such that $\lim_{n\to\infty} A^r z_n = \lim_{n\to\infty} S^r z_n = t$ for some $t \in X$ and any $r \in \mathbb{N}$, and $\lim_{n\to\infty} d(A^r S^r z_n, S^r A^r z_n) = 0$ for any $r \in \mathbb{N}$

Now we have the following example showing a pair of r-compatible mappings need not be conditionally r-compatible

> **Example 2.2.8**
>
> Let $X = [1, \infty)$ with Euclidean metric d. Define $A, S : X \mapsto X$ for any $r \in \mathbb{N}$ by $A^r x = x$ for all $x \in X$; $S^r x = 2^r x$ for all $x \in X$, then A and S are r-compatible but not conditionally r-compatible.

Now we have the following example showing a pair of conditionally r-compatible mappings can be r-compatible, depending upon certain condition on $r \in \mathbb{N}$

> **Example 2.2.9**
>
> Let $X = [1, 8]$ with the usual metric d. Define $A, S : X \mapsto X$, for any $r \in \mathbb{N}$ as follows. If r is even
> $$A^r x = \begin{cases} 2, & if\ x \leq 2 \\ 5, & if\ x > 2 \end{cases}$$
> and if r is odd
> $$A^r x = \begin{cases} 1, & if\ x \leq 2 \\ 5, & if\ x > 2 \end{cases}$$
> and if r is odd
> $$S^r x = \begin{cases} 3 - x, & if\ x \leq 2 \\ 8, & if\ x > 2 \end{cases}$$
> and if r is even
> $$S^r x = \begin{cases} x, & if\ x \leq 2 \\ 8, & if\ x > 2 \end{cases}$$
> Consider the constant sequence $z_n = 2$ and r is even, then $\lim_{n \to \infty} A^r z_n = \lim_{n \to \infty} S^r(z_n) = 2$, $\lim_{n \to \infty} A^r(S^r z_n) = \lim_{n \to \infty} S^r(A^r z_n) = 2$, and $\lim_{n \to \infty} d(A^r(S^r z_n), S^r(A^r z_n)) = 0$. Now consider the sequence $y_n = 2 - \frac{1}{n}$ and r is even, then $\lim_{n \to \infty} A^r y_n = \lim_{n \to \infty} S^r(y_n) = 2$. Also $\lim_{n \to \infty} A^r(S^r y_n) = \lim_{n \to \infty} S^r(A^r y_n) = 2$, and $\lim_{n \to \infty} d(A^r(S^r y_n), S^r(A^r y_n)) = 0$. So if r is even, then conditionally r-compatibilty of the mappings implies r-compatibility of them.
>
> Now consider r is odd, and the constant sequence $z_n = 2$, then $\lim_{n \to \infty} A^r z_n = \lim_{n \to \infty} S^r(z_n) = 1$, $1 = \lim_{n \to \infty} A^r(S^r z_n) \neq \lim_{n \to \infty} S^r(A^r z_n) = 2$, and $\lim_{n \to \infty} d(A^r(S^r z_n), S^r(A^r z_n)) \neq 0$. So if r is odd, the mappings are not conditionally r-compatible. Now consider the sequence $y_n = 2 - \frac{1}{n}$ and r is odd, then $\lim_{n \to \infty} A^r y_n = \lim_{n \to \infty} S^r(y_n) = 1$, $1 = \lim_{n \to \infty} A^r(S^r y_n) \neq \lim_{n \to \infty} S^r(A^r y_n) = 2$, and $\lim_{n \to \infty} d(A^r(S^r y_n), S^r(A^r y_n)) \neq 0$. So if r is odd, the mappings are neither conditionally r-compatible or r-compatible

Now we give the following example showing that conditional r-compatibility imply r-commutativity at the r-coincidence points.

> **Example 2.2.10**
>
> Let $X = [0, \infty)$ and d be the Euclidean metric on X. Define $A, S : X \mapsto X$ for any $r \in \mathbb{N}$ by $A^r(x) = x^{2^r}$ and
> $$S^r x = \begin{cases} x^{2^r}, & if\ x \in [0, 9] \cup (16, \infty) \\ x + 72r, & if\ x \in (9, 16] \end{cases}$$
> Now consider the sequence $y_n = \frac{1}{n}$, then $\lim_{n \to \infty} A^r y_n = 0$. On the other hand observe that $\lim_{n \to \infty} S^r y_n = 0$. Also $\lim_{n \to \infty} A^r(S^r y_n) = 0$ and $\lim_{n \to \infty} S^r(A^r y_n) = 0$, and $\lim_{n \to \infty} d(A^r(S^r y_n), S^r(A^r y_n)) = 0$. So A and S are conditionally r-compatible. Also observe for any $x \in [0, 9] \cup (16, \infty)$, $S^r(x) = A^r(x)$ and $A^r(S^r x) = S^r(A^r x)$, and so A and S r-commute at their r-coincidence points.

Inspired by faintly compatible mappings [Ravindra K Bisht and Naseer Shahzad, Faintly compatible mappings and common fixed points, Fixed Point Theory and Applications 2013, 2013:156] we introduce the notion of conditionally r-compatible maps, in a different way as follows

> **Definition 2.2.11**
>
> Two self-mappings A, S of a metric space X will be called faintly r-compatible iff A and S are conditionally r-compatible and A and S r-commute on a nonempty subset of r-coincidence points whenever the set of r-coincidences is nonempty

Now we give the following example showing faintly r-compatibility need not imply r-compatibility

> **Example 2.2.12**
>
> Let $X = [3, 6]$ and d be the usual metric on X. Define the self-mappings A and S for any $r \in \mathbb{N}$, on X as follows
>
> $$A^r x = \begin{cases} 3, & \text{if } x = 3 \text{ or } x > 5 \\ x + r, & \text{if } 3 < x \leq 5 \end{cases}$$
>
> and
>
> $$S^r x = \begin{cases} 3, & \text{if } x = 3 \\ x + 7r, & \text{if } 3 < x \leq 5 \\ x + r, & \text{if } x > 5 \end{cases}$$
>
> Consider the constant sequence $\{y_n := 3\}$, then $\lim_{n \to \infty} A^r(y_n) = \lim_{n \to \infty} S^r(y_n) = 3$, $\lim_{n \to \infty} S^r(A^r y_n) = \lim_{n \to \infty} A^r(S^r y_n) = 3$, and $\lim_{n \to \infty} d(S^r(A^r y_n), A^r(S^r y_n)) = 0$, thus A and S are conditionally r-compatible. Now observe $A^r x = S^r x$ iff $x = 3$. Moreover, $S^r(3) = A^r(3) = 3$ and $S^r(A^r(3)) = A^r(S^r(3)) = 3$, thus, A and S r-commute at their r-coincidence point. It now follows that A and S are faintly r-compatible. Now consider the sequence $\{z_n := 5 + \frac{1}{n}\}$, then, $3 = \lim_{n \to \infty} A^r(z_n) \neq \lim_{n \to \infty} S^r(z_n) = 5 + r$, $\lim_{n \to \infty} S^r(A^r z_n) = \lim_{n \to \infty} A^r(S^r z_n) = 3$, and $\lim_{n \to \infty} d(S^r(A^r z_n), A^r(S^r z_n)) = 0$, thus A and S are not r-compatible.

Now we give an example showing r-noncompatiblity does not imply faintly r-compatibility

> **Example 2.2.13**
>
> Let $X = [2, 10]$, and d be the usual metric on X. Define $A, S : X \mapsto X$ for any $r \in \mathbb{N}$ by
>
> $$A^r x = \begin{cases} 8, & \text{if } x \in [2, 5] \\ 2, & \text{if } x \in (5, 10] \end{cases}$$
>
> and
>
> $$S^r x = \begin{cases} 2, & \text{if } x \in [2, 5) \\ x - 3r, & \text{if } x \in [5, 10] \end{cases}$$
>
> Now consider $x_n = 5 + \frac{1}{n}$, then $2 = \lim_{n \to \infty} A^r x_n \neq \lim_{n \to \infty} S^r x_n = 5 - 3r$, and $\lim_{n \to \infty} S^r x_n = \lim_{n \to \infty} A^r x_n = 2$ iff $r = 1$, thus it is clear that the mappings are r-noncompatible. Now consider the constant sequence $\{z_n := 2\}$, observe that $8 = \lim_{n \to \infty} A^r z_n \neq \lim_{n \to \infty} S^r z_n = 2$. Thus, it follows that the mappings are not conditionally r-compatible, thus we can rule out faintly r-compatibility of the mappings.

Now we give an example showing faintly r-compatibility of the mappings implies r-compatibility of them

> **Example 2.2.14**
>
> Let $X = [1, \infty)$ and define $A, S : X \mapsto X$ for any $r \in \mathbb{N}$ and all $x \in X$ by $A^r x = x$ and $S^r x = 1$. Now consider the constant sequence $\{y_n := 1\}$, then $\lim_{n \to \infty} A^r y_n = \lim_{n \to \infty} S^r y_n = 1$, $\lim_{n \to \infty} S^r(A^r y_n) = A^r(S^r y_n) = 1$, and
>
> $$\lim_{n \to \infty} d(S^r(A^r y_n), A^r(S^r y_n)) = 0$$
>
> thus the mappings are conditionally r-compatible. Now $A^r x = S^r x$ iff $x = 1$, thus $x = 1$ is the only r-coincidence point. Moreover observe that $A^r(1) = S^r(1) = 1$ and $A^r(S^r(1)) = S^r(A^r(1)) = 1$, thus the mappings r-commute at their r-coincidence point. It now follows that the mappings are faintly r-compatible.
>
> Now consider $\{z_n := 1 + \frac{1}{n}\}$, $\lim_{n \to \infty} A^r z_n = \lim_{n \to \infty} S^r z_n = 1$, $\lim_{n \to \infty} S^r(A^r z_n) = A^r(S^r z_n) = 1$, and $\lim_{n \to \infty} d(S^r(A^r z_n), A^r(S^r z_n)) = 0$. It follows that the mappings are r-compatible.

2.3 Main Results

> **Definition 2.3.1**
>
> Let (X, d) be a metric space, and $A, S : X \mapsto X$. We say A is a Banach contraction with respect to S or a Jungck contraction if for all $x, y \in X$ and $k \in [0, 1)$ we have $d(Ax, Ay) \leq k d(Sx, Sy)$

> **Remark 2.3.2**
>
> Observe that if S is the identity in the above definition, then A is a Banach contraction

> **Definition 2.3.3**
>
> Let (X, d) be a metric space, and $A, S : X \mapsto X$. We say A is a higher-order Banach contraction with respect to S or a higher-order Jungck contraction if for all $x, y \in X$ the following holds
>
> $$d(A^r x, A^r y) \leq \sum_{q=0}^{r-1} c_q d(S^{q+1} x, S^{q+1} y)$$
>
> where $0 \leq c_q < 1$ for all $0 \leq q \leq r - 1$ and $r \in \mathbb{N}$

> **Proposition 2.3.4**
>
> Let (X, d) be a metric space, and $A, S : X \mapsto X$, such that A is a higher-order Banach contraction with respect to S. For every pair $x \neq y$, define
>
> $$Z := Z(x, y) = \max_{0 \leq v \leq r-1} \beta^{-v} \frac{d(A^v x, A^v y)}{d(Sx, Sy)}$$
>
> then
>
> $$Z = \max_{n \in \mathbb{N} \cup \{0\}} \beta^{-n} \frac{d(A^n x, A^n y)}{d(Sx, Sy)}$$
>
> where $\beta \in [0, 1)$

Now by the previous proposition, we have the following

Definition 2.3.5

Let (X, d) be a metric space, and $A, S : X \mapsto X$. We say A is a higher-order Banach contraction with respect to S or a higher-order Jungck contraction if for all $x, y \in X$ the following holds

$$d(A^r x, A^r y) \leq Z\beta^r d(Sx, Sy)$$

where $Z \geq 1$ is given by the previous proposition and $\beta \in [0, 1)$

Now we prove the following

Theorem 2.3.6

Let A and S be r-noncompatible faintly r-compatible self-mappings of a metric space (X, d) satisfying

(a) $A^r(X) \subseteq S^r(X)$ for any $r \in \mathbb{N}$

(b) Definition 2.3.5

If either A or S is r-continuous, then A and S have a unique common r-fixed point.

Proof of Theorem 2.3.6

r-Noncompatibility of A and S implies there exists a sequence $\{x_n\}$ in X such that $\lim_{n \to \infty} A^r x_n = \lim_{n \to \infty} S^r x_n = t$ for some $t \in X$, but $\lim_{n \to \infty} d(A^r(S^r x_n), S^r(A^r x_n))$ is either nonzero or doesn't exist. Since A and S are faintly r-compatible and $\lim_{n \to \infty} A^r x_n = \lim_{n \to \infty} S^r x_n = t$, there exists a sequence $\{z_n\} \in X$ satisfying $\lim_{n \to \infty} A^r z_n = \lim_{n \to \infty} S^r z_n = u$(say) such that $\lim_{n \to \infty} d(A^r(S^r z_n), S^r(A^r z_n)) = 0$. Further since A is r-continuous, then $\lim_{n \to \infty} A^r(A^r z_n) = A^r u$ and $\lim_{n \to \infty} A^r(S^r z_n) = A^r u$. The last three limits together imply $\lim_{n \to \infty} S^r(A^r z_n) = A^r u$. Now observe $A^r(X) \subseteq S^r(X)$ implies that $A^r u = S^r v$ for some $v \in X$ and $A^r(A^r z_n) \to S^r v$, $S^r(A^r z_n) \to S^r v$. By Definition 2.3.5,

$$d(A^r v, A^r(A^r z_n)) \leq Z\beta^r d(S^r v, S^r(A^r z_n))$$

thus taking limits in the above inequality as $n \to \infty$ we deduce that $A^r v = S^r v$. It follows that v is a r-coincidence point of A and S. Further, faint r-compatibility implies $A^r(S^r(v)) = S^r(A^r v)$, and hence $A^r(S^r(v)) = S^r(A^r v) = A^r(A^r v) = S^r(S^r v)$. If $A^r(v) \neq A^r(A^r v)$, then Definition 2.3.5 implies

$$d(A^r v, A^r(A^r v)) \leq Z\beta^r d(S^r v, S^r(A^r v)) = Z\beta^r d(A^r v, A^r(A^r v))$$

which is a contradiction. Hence $A^r v$ is a common r-fixed point of A and S. The same conclusion is obtained when S is assumed to be r-continuous, since the r-continuity of S implies the r-continuity of A. The uniqueness of the common r-fixed point follows from Definition 2.3.5 , and the proof is finished.

2.4 Open Problems

Problem 2.4.1

Recall the concept of Kannan mapping appeared in [R. Kannan, Some results on fixed points, Bull. Calcutta Math. Soc. 60(1968), 71-76.]

Definition 2.4.1.1

Let (X, d) be a metric space, and $A, S : X \mapsto X$. We say A is a Kannan contraction with respect to S or a Jungck-Kannan contraction if for all $x, y \in X$ and $k \in [0, \frac{1}{2})$ we have

$$d(Ax, Ay) \leq k[d(Sx, Ax) + d(Sy, Ay)]$$

> **Remark 2.4.1.2**
> Observe that if S is the identity in the above definition, then A is a Kannan contraction

> **Definition 2.4.1.3**
> Let (X, d) be a metric space, and $A, S : X \mapsto X$. We say A is a higher-order Kannan contraction with respect to S or a higher-order Jungck-Kannan contraction if for all $x, y \in X$ the following holds
> $$d(A^r x, A^r y) \leq \sum_{q=0}^{r-1} c_q [d(S^{q+1} x, A^{q+1} x) + d(S^{q+1} y, A^{q+1} y)]$$
> where $0 \leq c_q < \frac{1}{2}$ for all $0 \leq q \leq r-1$ and $r \in \mathbb{N}$

> **Proposition 2.4.1.4**
> Let (X, d) be a metric space, and $A, S : X \mapsto X$, such that A is a higher-order Kannan contraction with respect to S. For every pair $x \neq y$, define
> $$Z := Z(x, y) = \max_{0 \leq v \leq r-1} \beta^{-v} \frac{d(A^v x, A^v y)}{d(Sx, Ax) + d(Sy, Ay)}$$
> then
> $$Z = \max_{n \in \mathbb{N} \cup \{0\}} \beta^{-n} \frac{d(A^n x, A^n y)}{d(Sx, Ax) + d(Sy, Ay)}$$
> where $\beta \in [0, \frac{1}{2})$

Now by the previous proposition, we have the following

> **Definition 2.4.1.5**
> Let (X, d) be a metric space, and $A, S : X \mapsto X$. We say A is a higher-order Kannan contraction with respect to S or a higher-order Jungck-Kannan contraction if for all $x, y \in X$ the following holds
> $$d(A^r x, A^r y) \leq Z \beta^r [d(Sx, Ax) + d(Sy, Ay)]$$
> where $Z \geq 1$ is given by the previous proposition and $\beta \in [0, \frac{1}{2})$

Now the open problem is to prove the following

> **Theorem 2.4.1.6**
> Let A and S be r-noncompatible faintly r-compatible self-mappings of a metric space (X, d) satisfying
>
> (a) $A^r(X) \subseteq S^r(X)$ for any $r \in \mathbb{N}$
>
> (b) Definition 2.4.1.5
>
> If either A or S is r-continuous, then A and S have a unique common r-fixed point.

Problem 2.4.2

Recall the concept of Chatterjea mapping appeared in [S. K. Chatterjea, "Fixed-point theorems," Comptes Rendus de l'Academie Bulgare des Sciences, vol. 25, pp. 727–730, 1972]. Now we introduce the following

Definition 2.4.2.1

Let (X,d) be a metric space, and $A, S : X \mapsto X$. We say A is a Chatterjea contraction with respect to S or a Jungck-Chatterjea contraction if for all $x, y \in X$ and $k \in [0, \frac{1}{2})$ we have

$$d(Ax, Ay) \leq k[d(Sy, Ax) + d(Sx, Ay)]$$

Remark 2.4.2.2

Observe that if S is the identity in the above definition, then A is a Chatterjea contraction

Definition 2.4.2.3

Let (X,d) be a metric space, and $A, S : X \mapsto X$. We say A is a higher-order Chatterjea contraction with respect to S or a higher-order Jungck-Chatterjea contraction if for all $x, y \in X$ the following holds

$$d(A^r x, A^r y) \leq \sum_{q=0}^{r-1} c_q [d(S^{q+1}y, A^{q+1}x) + d(S^{q+1}x, A^{q+1}y)]$$

where $0 \leq c_q < \frac{1}{2}$ for all $0 \leq q \leq r-1$ and $r \in \mathbb{N}$

Proposition 2.4.2.4

Let (X,d) be a metric space, and $A, S : X \mapsto X$, such that A is a higher-order Chatterjea contraction with respect to S. For every pair $x \neq y$, define

$$Z := Z(x,y) = \max_{0 \leq v \leq r-1} \beta^{-v} \frac{d(A^v x, A^v y)}{d(Sy, Ax) + d(Sx, Ay)}$$

then

$$Z = \max_{n \in \mathbb{N} \cup \{0\}} \beta^{-n} \frac{d(A^n x, A^n y)}{d(Sy, Ax) + d(Sx, Ay)}$$

where $\beta \in [0, \frac{1}{2})$

Now by the previous proposition, we have the following

Definition 2.4.2.5

Let (X,d) be a metric space, and $A, S : X \mapsto X$. We say A is a higher-order Chatterjea contraction with respect to S or a higher-order Jungck-Chatterjea contraction if for all $x, y \in X$ the following holds

$$d(A^r x, A^r y) \leq Z \beta^r [d(Sy, Ax) + d(Sx, Ay)]$$

where $Z \geq 1$ is given by the previous proposition and $\beta \in [0, \frac{1}{2})$

Now the open problem is to prove the following

Theorem 2.4.2.6

Let A and S be r-noncompatible faintly r-compatible self-mappings of a metric space (X,d) satisfying

(a) $A^r(X) \subseteq S^r(X)$ for any $r \in \mathbb{N}$

(b) Definition 2.4.2.5

If either A or S is r-continuous, then A and S have a unique common r-fixed point.

Problem 2.4.3

> **Definition 2.4.3.1**
>
> Let T and A be two self-maps of a metric space (X, d) into itself. We say T is Hardy and Rogers type contraction with respect to A if the following holds for all $x, y \in X$ and $0 \leq k < \frac{1}{6}$
>
> $$d(Tx, Ty) \leq k[d(Ax, Ay) + d(Ax, Tx) + d(Ay, Ty) + d(Ax, Ty) + d(Ay, Tx)]$$

> **Definition 2.4.3.2**
>
> Let T and A be two self-maps of a metric space (X, d) into itself. We say T is a higher-order Hardy and Rogers type contraction with respect to A if the following holds for all $x, y \in X$
>
> $$d(T^r x, T^r y) \leq \sum_{q=0}^{r-1} c_q [d(A^{q+1} x, A^{q+1} y) + d(A^{q+1} x, T^{q+1} x) + d(A^{q+1} y, T^{q+1} y)$$
> $$+ d(A^{q+1} x, T^{q+1} y) + d(A^{q+1} y, T^{q+1} x)]$$
>
> where $0 \leq c_q < \frac{1}{6}$, for all $0 \leq q \leq r-1$, and $r \in \mathbb{N}$

> **Proposition 2.4.3.3**
>
> Let (X, d) be metric space, and $T, A : X \mapsto X$, such that T is a higher-order Hardy and Rogers type contraction with respect to A. For every pair $x \neq y$, define
>
> $$Z := Z(x, y) = \max_{0 \leq v \leq r-1} \beta^{-v} \frac{d(T^v x, T^v y)}{d(Ax, Ay) + d(Ax, Tx) + d(Ay, Ty) + d(Ax, Ty) + d(Ay, Tx)}$$
>
> then
>
> $$Z = \max_{n \in \mathbb{N} \cup \{0\}} \beta^{-n} \frac{d(T^n x, T^n y)}{d(Ax, Ay) + d(Ax, Tx) + d(Ay, Ty) + d(Ax, Ty) + d(Ay, Tx)}$$
>
> where $\beta \in [0, \frac{1}{6})$

Now using the above Proposition, we have the following alternate characterization of Definition 2.4.3.2

> **Definition 2.4.3.4**
>
> Let T and A be two self-maps of a metric space (X, d) into itself. We say the pair T is higher-order Hardy and Rogers type contraction with respect to A if the following holds for all $x, y \in X$ and $r \in \mathbb{N}$
>
> $$d(T^r x, T^r y) \leq Z \beta^r [d(Ax, Ay) + d(Ax, Tx) + d(Ay, Ty) + d(Ax, Ty) + d(Ay, Tx)]$$
>
> where $Z \geq 1$ is given by the previous Proposition, and $\beta \in [0, \frac{1}{6})$

Now the open problem is to prove the following

> **Theorem 2.4.3.5**
>
> Let T and A be r-noncompatible faintly r-compatible self-mappings of a metric space (X, d) satisfying
>
> (a) $T^r(X) \subseteq A^r(X)$ for any $r \in \mathbb{N}$
>
> (b) Definition 2.4.3.4
>
> If either T or A is r-continuous, then T and A have a unique common r-fixed point.

Chapter 3

A Higher-Order Common Fixed Point Theorem under r-Compatible Mappings of Type (R) in Metric Space

3.1 Brief Summary

> **Abstract**
>
> Inspired by higher-order fixed point theory [Clement Ampadu, Fixed Point Theory for Higher-Order Mappings. ISBN: 5800118959925, lulu.com, 2016] and compatible mappings of type (R) [Y. Rohen, M.R. Singh and L. Shambhu, Common fixed points of compatible mapping of type (C) in Banach Spaces, Proc. of Math. Soc., BHU 20(2004), 77-87] we introduce higher-order compatible mappings of type (R), and obtain the higher-order version of Theorem 2.4 [M. Koireng Meitei, Leenthoi Ningombam and Yumnam Rohen, Common Fixed Points of Compatible Mappings of Type (R), Gen. Math. Notes, Vol. 10, No. 1, May 2012, pp. 58-62] in the setting of metric spaces.

3.2 Preliminaries

Inspired by [G. Jungck, Compatible maps and common fixed points, Inter .J. Math. and Math. Sci., 9(1986), 771-779] we introduce the following

> **Definition 3.2.1**
>
> Let S and T be mappings from a complete metric space X into itself. We say the mappings S and T are r-compatible if $\lim_{n\to\infty} d(S^r T^r x_n, T^r S^r x_n) = 0$, whenever $\{x_n\}$ is a sequence in X such that $\lim_{n\to\infty} S^r x_n = T^r x_n = t$ for some $t \in X$ and any $r \in \mathbb{N}$

Inspired by [H.K. Pathak, S.S. Chang and Y.J. Cho., Fixed point theorem for compatible mappings of type (P), Indian J. Math. 36(2) (1994), 151-166] we introduce the following

> **Definition 3.2.2**
>
> Let S and T be mappings from a complete metric space X into itself. We say the mappings S and T are r-compatible of type (P) if $\lim_{n\to\infty} d(S^r S^r x_n, T^r T^r x_n) = 0$, whenever $\{x_n\}$ is a sequence in X such that $\lim_{n\to\infty} S^r x_n = T^r x_n = t$ for some $t \in X$ and any $r \in \mathbb{N}$

Inspired by [Y. Rohen, M.R. Singh and L. Shambhu, Common fixed points of compatible mapping of type (C) in Banach Spaces, Proc. of Math. Soc., BHU 20(2004), 77-87] we introduce the following

> **Definition 3.2.3**
>
> Let S and T be mappings from a complete metric space X into itself. We say the mappings S and T are r-compatible of type (R) if $\lim_{n\to\infty} d(S^r T^r x_n, T^r S^r x_n) = 0$ and $\lim_{n\to\infty} d(S^r S^r x_n, T^r T^r x_n) = 0$, whenever $\{x_n\}$ is a sequence in X such that $\lim_{n\to\infty} S^r x_n = T^r x_n = t$ for some $t \in X$ and any $r \in \mathbb{N}$

3.3 Main Results

In the sequel we will need the following Propositions whose metric counterpart when $r = 1$ can be found in [Y. Rohen, M.R. Singh and L. Shambhu, Common fixed points of compatible mapping of type (C) in Banach Spaces, Proc. of Math. Soc., BHU 20(2004), 77-87]

> **Proposition 3.3.1**
>
> Let S and T be mappings from a complete metric space (X, d) into itself. If the pair (S, T) is r-compatible of type (R) on X and $S^r z = T^r z$ for any $r \in \mathbb{N}$ and $z \in X$ then
>
> $$S^r T^r z = T^r S^r z = S^r S^r z = T^r T^r z$$

> **Proposition 3.3.2**
>
> Let S and T be mappings from a complete metric space (X, d) into itself. If a pair (S, T) is r-compatible of type (R) on X and $\lim_{n\to\infty} S^r x_n = \lim_{n\to\infty} T^r x_n = z$ for some $z \in X$ and any $r \in \mathbb{N}$, then we have
>
> (a) $\lim_{n\to\infty} d(T^r S^r x_n, S^r z) = 0$, if S is r-continuous
>
> (b) $\lim_{n\to\infty} d(T^r S^r x_n, T^r z) = 0$, if T is r-continuous
>
> (c) $S^r T^r z = T^r S^r z$ and $T^r z = S^r z$ if S and T are r-continuous at z

Inspired by the mapping contained in [M. Koireng Meitei, Leenthoi Ningombam and Yumnam Rohen, Common Fixed Points of Compatible Mappings of Type (R), Gen. Math. Notes, Vol. 10, No. 1, May 2012, pp. 58-62] we introduce the following

> **Definition 3.3.3**
>
> Let A, B, S, T be four mappings from a metric space (X, d) into itself. We say (A, B) is an MNR-type contraction with respect to (S, T) if there exists $k \in [0, \frac{1}{3})$ such that the following holds for all $x, y \in X$
>
> $$\begin{aligned} d(Ax, By)^2 \leq &\, k[d(Ax, Sx)d(By, Ty) + d(By, Sx)d(Ax, Ty) + d(Ax, Sx)d(Ax, Ty) \\ &+ d(By, Ty)d(By, Sx)] \end{aligned}$$

The higher-order version of the above reads as follows

CHAPTER 3. A HIGHER-ORDER COMMON FIXED POINT THEOREM UNDER R-COMPATIBLE MAPPINGS OF TYPE (R) IN METRIC SPACE

Definition 3.3.4

Let A, B, S, T be four mappings from a metric space (X, d) into itself. We say (A, B) is a higher-order MNR-type contraction with respect to (S, T) if there exists $c_q \in [0, \frac{1}{3})$ such that for all $0 \leq q \leq r - 1$ and $r \in \mathbb{N}$, the following holds for all $x, y \in X$

$$d(A^r x, B^r y)^2 \leq \sum_{q=0}^{r-1} \Big\{ c_q [d(A^{q+1} x, S^{q+1} x) d(B^{q+1} y, T^{q+1} y)$$
$$+ d(B^{q+1} y, S^{q+1} x) d(A^{q+1} x, T^{q+1} y) + d(A^{q+1} x, S^{q+1} x) d(A^{q+1} x, T^{q+1} y)$$
$$+ d(B^{q+1} y, T^{q+1} y) d(B^{q+1} y, S^{q+1} x)] \Big\}$$

Now we introduce the following which allows an alternate characterization of the higher-order MNR-type contraction

Proposition 3.3.5

Let A, B, S, T be four mappings from a metric space (X, d) into itself, where (A, B) is a higher-order MNR-type contraction with respect to (S, T). Put

$$ABST(x, y) := d(Ax, Sx) d(By, Ty) + d(By, Sx) d(Ax, Ty) + d(Ax, Sx) d(Ax, Ty)$$
$$+ d(By, Ty) d(By, Sx)$$

Now for every pair $x \neq y$, define

$$Z := Z(x, y) = \max_{0 \leq v \leq r-1} \beta^{-v} \frac{d(A^v x, B^v y)^2}{ABST(x, y)}$$

then

$$Z = \max_{n \in \mathbb{N} \cup \{0\}} \beta^{-n} \frac{d(A^n x, B^n y)^2}{ABST(x, y)}$$

where $\beta \in [0, \frac{1}{3})$.

Now by the above Proposition, we have the following alternate characterization of the higher-order MNR-type contraction

Definition 3.3.6

Let A, B, S, T be four mappings from a metric space (X, d) into itself. We say (A, B) is a higher-order MNR-type contraction with respect to (S, T) if the following holds for all $x, y \in X$ and any $r \in \mathbb{N}$

$$d(A^r x, B^r y)^2 \leq Z \beta^r [d(Ax, Sx) d(By, Ty) + d(By, Sx) d(Ax, Ty) + d(Ax, Sx) d(Ax, Ty)$$
$$+ d(By, Ty) d(By, Sx)]$$

where $Z \geq 1$ is given by the previous Proposition and $\beta \in [0, \frac{1}{3})$

In order to prove the main result we need the following

Lemma 3.3.7

Let A, B, S, T be four mappings from a metric space (X, d) into itself, satisfying the following conditions

(a) $A^r(X) \subseteq T^r(X)$ and $B^r(X) \subseteq S^r(X)$ for any $r \in \mathbb{N}$

(b) Definition 3.3.6

then the sequence $\{y_n\}$, for any $r \in \mathbb{N}$, defined by

$$y_{2n+1} = T^r x_{2n+1} = A^r x_{2n} \text{ and } y_{2n} = S^r x_{2n} = B^r x_{2n-1}$$

is a Cauchy sequence in X

Proof of Lemma 3.3.7

By the definition of $\{y_n\}$ and Definition 3.3.6, we deduce the following

$$\begin{aligned}
d(y_{2n+1}, y_{2n})^2 &= d(A^r x_{2n}, B^r x_{2n-1})^2 \\
&\leq Z\beta^r [d(A^r x_{2n}, S^r x_{2n}) d(B^r x_{2n-1}, T^r x_{2n-1}) \\
&\quad + d(B^r x_{2n-1}, S^r x_{2n}) d(A^r x_{2n}, T^r x_{2n-1}) \\
&\quad + d(A^r x_{2n}, S^r x_{2n}) d(A^r x_{2n}, T^r x_{2n-1}) \\
&\quad + d(B^r x_{2n-1}, T^r x_{2n-1}) d(B^r x_{2n-1}, S^r x_{2n})] \\
&= Z\beta^r [d(y_{2n+1}, y_{2n}) d(y_{2n}, y_{2n-1}) \\
&\quad + d(y_{2n}, y_{2n}) d(y_{2n+1}, y_{2n-1}) + d(y_{2n+1}, y_{2n}) d(y_{2n+1}, y_{2n-1}) \\
&\quad + d(y_{2n}, y_{2n-1}) d(y_{2n}, y_{2n})] \\
&= Z\beta^r [d(y_{2n+1}, y_{2n}) d(y_{2n}, y_{2n-1}) + d(y_{2n+1}, y_{2n}) d(y_{2n+1}, y_{2n-1})] \\
&\leq Z\beta^r [d(y_{2n+1}, y_{2n}) d(y_{2n}, y_{2n-1}) + d(y_{2n+1}, y_{2n})^2 \\
&\quad + d(y_{2n+1}, y_{2n}) d(y_{2n}, y_{2n-1})] \\
&= Z\beta^r [2 d(y_{2n+1}, y_{2n}) d(y_{2n}, y_{2n-1}) + d(y_{2n+1}, y_{2n})^2] \\
&\leq Z\beta^r [2 d(y_{2n}, y_{2n-1})^2 + d(y_{2n+1}, y_{2n})^2]
\end{aligned}$$

From the above we deduce that

$$d(y_{2n+1}, y_{2n})^2 \leq \frac{2Z\beta^r}{1 - Z\beta^r} d(y_{2n}, y_{2n-1})^2$$

or equivalently

$$d(y_{2n+1}, y_{2n}) \leq \sqrt{\frac{2Z\beta^r}{1 - Z\beta^r}} d(y_{2n}, y_{2n-1})$$

Since $\frac{2Z\beta^r}{1-Z\beta^r} < 1$, consequently the sequence $\{y_n\}$ is Cauchy, and the proof is finished

Now our main result is the following

Theorem 3.3.8

Let A, B, S, T be four mappings from a complete metric space (X, d) into itself, satisfying the following conditions

(a) $A^r(X) \subseteq T^r(X)$ and $B^r(X) \subseteq S^r(X)$ for any $r \in \mathbb{N}$

(b) Definition 3.3.6

(c) One of A, B, S, T is r-continuous

(d) The pair (A, S) and (B, T) are r-compatible of type (R) on X

Then A, B, S, T have a unique common r-fixed point in X

CHAPTER 3. A HIGHER-ORDER COMMON FIXED POINT THEOREM UNDER R-COMPATIBLE MAPPINGS OF TYPE (R) IN METRIC SPACE

Proof of Theorem 3.3.8

Let $x_0 \in X$, then since $A^r(X) \subseteq T^r(X)$ and $B^r(X) \subseteq S^r(X)$ for any $r \in \mathbb{N}$, we can construct a sequence $\{y_n\} \in X$, for any $r \in \mathbb{N}$, by

$$y_{2n+1} = T^r x_{2n+1} = A^r x_{2n} \text{ and } y_{2n} = S^r x_{2n} = B^r x_{2n-1}$$

and by Lemma 3.3.7 this sequence in Cauchy in X. Since X is complete, there is $z \in X$ such that $\lim_{n \to \infty} y_n = z$. Consequently, the sub-sequences $\{A^r x_{2n}\}$, $\{S^r x_{2n}\}$, $\{B^r x_{2n-1}\}$, and $\{T^r x_{2n+1}\}$, also converge to z. Suppose S is r-continuous. Since A and S are r-compatible of type (R) on X, then by Proposition 3.3.2, we have $\lim_{n \to \infty} S^{2r} x_{2n} = S^r z$ and $\lim_{n \to \infty} A^r S^r x_{2n} = S^r z$. Now by Definition 3.3.6, observe we have the following

$$\begin{aligned}d(A^r S^r x_{2n}, B^r x_{2n-1})^2 &\leq Z\beta^r [d(A^r S^r x_{2n}, S^{2r} x_{2n})d(B^r x_{2n-1}, T^r x_{2n-1}) \\&+ d(B^r x_{2n-1}, S^{2r} x_{2n})d(A^r S^r x_{2n-1}, T^r x_{2n-1}) \\&+ d(A^r S^r x_{2n}, S^{2r} x_{2n})d(A^r S^r x_{2n}, T^r x_{2n-1}) \\&+ d(B^r x_{2n-1}, T^r x_{2n-1})d(B^r x_{2n-1}, S^{2r} x_{2n})]\end{aligned}$$

Now taking limits in the above as $n \to \infty$, we deduce that

$$d(S^r z, z)^2 \leq Z\beta^r d(S^r z, z)^2$$

which is a contradiction. Hence $S^r z = z$. Now observe we have the following

$$\begin{aligned}d(A^r z, B^r x_{2n-1})^2 &\leq Z\beta^r [d(A^r z, S^r z)d(B^r x_{2n-1}, T^r x_{2n-1}) \\&+ d(B^r x_{2n-1}, S^r z)d(A^r z, T^r x_{2n-1}) \\&+ d(A^r z, S^r z)d(A^r z, T^r x_{2n-1}) \\&+ d(B^r x_{2n-1}, T^r x_{2n-1})d(B^r x_{2n-1}, S^r z)]\end{aligned}$$

Now taking limits in the above as $n \to \infty$, we deduce that

$$d(A^r z, z)^2 \leq Z\beta^r d(A^r z, z)^2$$

which is a contradiction. Hence $A^r z = z$, and by part (a) of the Theorem, $z \in T^r(X)$. Now since T is a self-map of X there is a point $u \in X$ such that $z = A^r z = T^r u$. Now observe we have the following from Definition 3.3.6

$$\begin{aligned}d(z, B^r u)^2 &= d(A^r z, B^r u)^2 \\&\leq Z\beta^r [d(A^r z, S^r z)d(B^r u, T^r u) \\&+ d(B^r u, S^r z)d(A^r z, T^r u) \\&+ d(A^r z, S^r z)d(A^r z, T^r u) \\&+ d(B^r u, T^r u)d(B^r u, S^r z)]\end{aligned}$$

From the above, we deduce that $d(z, B^r u)^2 \leq Z\beta^r d(z, B^r u)^2$, which is a contradiction. Hence $z = B^r u$. It follows that $z = T^r u = B^r u$. Since the pair (A, S) and (B, T) are r-compatible of type (R) on X, we have $d(T^r B^r u, B^r T^r u) = 0$, that is, $d(T^r z, B^r z) = 0$, and hence $T^r z = B^r z$.

> **Proof of Theorem 3.3.8 Continued**
>
> Now observe we have the following from Definition 3.3.6
>
> $$\begin{aligned}d(z,T^r z)^2 &= d(A^r z, B^r z)^2 \\ &\leq Z\beta^r[d(A^r z, S^r z)d(B^r z, T^r z) \\ &\quad + d(B^r z, S^r z)d(A^r z, T^r z) \\ &\quad + d(A^r z, S^r z)d(A^r z, T^r z) \\ &\quad + d(B^r z, T^r z)d(B^r z, S^r z)]\end{aligned}$$
>
> From the above we deduce that $d(z,T^r z)^2 \leq Z\beta^r d(z,T^r z)^2$, which is a contradiction. Hence $z = T^r z$. It follows that $z = T^r z = B^r z$. Thus, it is now clear that z is a common r-fixed point of A, B, S, T. If any one of A, B, or T is r-continuous, the result still holds. Finally we show uniqueness of the common r-fixed point. Suppose w is another common r-fixed point of A, B, S, T, then observe from Definition 3.3.6, we have the following
>
> $$\begin{aligned}d(z,w)^2 &= d(A^r z, B^r w)^2 \\ &\leq Z\beta^r[d(A^r z, S^r z)d(B^r w, T^r w) \\ &\quad + d(B^r w, S^r z)d(A^r z, T^r w) \\ &\quad + d(A^r z, S^r z)d(A^r z, T^r w) \\ &\quad + d(B^r w, T^r w)d(B^r w, S^r z)]\end{aligned}$$
>
> From the above we deduce that $d(z,w)^2 \leq Z\beta^r d(z,w)^2$, which is a contradiction. Hence $z = w$

3.4 Open Problem

Problem 3.4.1

Compatible mappings of type (P) have appeared in the literature, and for example, see [V. Srinivas and V. Naga Raju, Common Fixed Point Theorem on Compatible Mappings of Type (P), Gen. Math. Notes, Vol. 21, No. 2, April 2014, pp. 87-94]. Now we introduce the following

> **Definition 3.4.1.1**
>
> Let S and T be mappings from a metric space X into itself. The mappings S and T will be called r-compatible mappings of type (P) if $\lim_{n\to\infty} d(S^r S^r x_n, T^r T^r x_n) = 0$ whenever $\{x_n\}$ is a sequence in X such that $\lim_{n\to\infty} S^r x_n = \lim_{n\to\infty} T^r x_n = t$ for some $t \in X$ and any $r \in \mathbb{N}$

The open problem is to prove the following

> **Theorem 3.4.1.2**
>
> Let A, B, S, T be four mappings from a complete metric space (X,d) into itself, satisfying the following conditions
>
> (a) $A^r(X) \subseteq T^r(X)$ and $B^r(X) \subseteq S^r(X)$ for any $r \in \mathbb{N}$
>
> (b) Definition 3.3.6
>
> (c) One of A, B, S, T is r-continuous
>
> (d) The pair (A,S) and (B,T) are r-compatible of type (P) on X
>
> Then A, B, S, T have a unique common r-fixed point in X

Chapter 4

A Common Higher-Order Fixed Point Theorem under r-Compatible Mappings of Type (P) in Metric Space

4.1 Brief Summary

> **Abstract**
>
> Inspired by higher-order fixed point theory [Clement Ampadu, Fixed Point Theory for Higher-Order Mappings. ISBN: 5800118959925, lulu.com, 2016], we introduce r-compatible mappings of type (P), and obtain the higher-order version of Theorem 3.1 [V. Srinivas and V. Naga Raju, Common Fixed Point Theorem on Compatible Mappings of Type (P), Gen. Math. Notes, Vol. 21, No. 2, April 2014, pp. 87-94]

4.2 Preliminaries

Inspired by [G. Jungck, Compatible maps and common fixed points, Inter .J. Math. and Math. Sci., 9(1986), 771-779] we introduce the following

> **Definition 4.2.1**
>
> Let S and T be mappings from a complete metric space X into itself. We say the mappings S and T are r-compatible if $\lim_{n\to\infty} d(S^r T^r x_n, T^r S^r x_n) = 0$, whenever $\{x_n\}$ is a sequence in X such that $\lim_{n\to\infty} S^r x_n = T^r x_n = t$ for some $t \in X$ and any $r \in \mathbb{N}$

Taking inspiration from [G. Jungck, P.P. Murthy and Y.J. Cho, Compatible mappings of type (A) and common fixed points, Math. Japonica, 38(1993), 381-390] we introduce the following

> **Definition 4.2.2**
>
> Let S and T be mappings from a metric space X into itself. The mappings S and T will be called r-compatible mappings of type (A) if $\lim_{n\to\infty} d(S^r T^r x_n, T^r T^r x_n) = 0$ and $\lim_{n\to\infty} d(T^r S^r x_n, S^r S^r x_n) = 0$ whenever $\{x_n\}$ is a sequence in X such that $\lim_{n\to\infty} S^r x_n = \lim_{n\to\infty} T^r x_n = z$ for some $z \in X$

Compatible mappings of type (B) have appeared in the literature, and for example, see [V. Srinivas and V. Naga Raju, Common Fixed Point Theorem on Compatible Mappings of Type (P), Gen. Math. Notes, Vol. 21, No. 2, April 2014, pp. 87-94]. Now we introduce the following

CHAPTER 4. A COMMON HIGHER-ORDER FIXED POINT THEOREM UNDER R-COMPATIBLE MAPPINGS OF TYPE (P) IN METRIC SPACE

> **Definition 4.2.3**
>
> Let S and T be mappings from a metric space X into itself. The mappings S and T will be called r-compatible mappings of type (B) if
>
> $$\lim_{n\to\infty} d(S^r T^r x_n, T^r T^r x_n) \leq \frac{1}{2}[\lim_{n\to\infty} d(S^r T^r x_n, S^r t) + \lim_{n\to\infty} d(S^r t, S^r S^r x_n)]$$
>
> and
>
> $$\lim_{n\to\infty} d(T^r S^r x_n, S^r S^r x_n) \leq \frac{1}{2}[\lim_{n\to\infty} d(T^r S^r x_n, T^r t) + \lim_{n\to\infty} d(T^r t, T^r T^r x_n)]$$
>
> whenever $\{x_n\}$ is a sequence in X such that $\lim_{n\to\infty} S^r x_n = \lim_{n\to\infty} T^r x_n = t$ for some $t \in X$

Compatible mappings of type (P) have appeared in the literature, and for example, see [V. Srinivas and V. Naga Raju, Common Fixed Point Theorem on Compatible Mappings of Type (P), Gen. Math. Notes, Vol. 21, No. 2, April 2014, pp. 87-94]. Now we introduce the following

> **Definition 4.2.4**
>
> Let S and T be mappings from a metric space X into itself. The mappings S and T will be called r-compatible mappings of type (P) if $\lim_{n\to\infty} d(S^r S^r x_n, T^r T^r x_n) = 0$ whenever $\{x_n\}$ is a sequence in X such that $\lim_{n\to\infty} S^r x_n = \lim_{n\to\infty} T^r x_n = t$ for some $t \in X$

Now inspired by the mapping in Theorem 2.5 of [V. Srinivas and V. Naga Raju, Common Fixed Point Theorem on Compatible Mappings of Type (P), Gen. Math. Notes, Vol. 21, No. 2, April 2014, pp. 87-94] we introduce the following

> **Definition 4.2.5**
>
> Let A, B, S, T be self mappings from a metric space (X, d) into itself. We say the pair (A, B) is an SC-type contraction with respect to (S, T) if the following holds for all $x, y \in X$ and $k \in [0, \frac{1}{3})$
>
> $$d(Ax, By)^2 \leq k[d(Ax, Sx)d(By, Ty) + d(By, Sx)d(Ax, Ty) + d(Ax, Sx)d(Ax, Ty) \\ + d(By, Ty)d(By, Sx)]$$

Now we introduce the higher-order version of the above as follows

> **Definition 4.2.6**
>
> Let A, B, S, T be four mappings from a metric space (X, d) into itself. We say (A, B) is a higher-order SC-type contraction with respect to (S, T) if there exists $c_q \in [0, \frac{1}{3})$ such that for all $0 \leq q \leq r-1$ and $r \in \mathbb{N}$, the following holds for all $x, y \in X$
>
> $$d(A^r x, B^r y)^2 \leq \sum_{q=0}^{r-1} \Big\{ c_q[d(A^{q+1}x, S^{q+1}x)d(B^{q+1}y, T^{q+1}y) \\ + d(B^{q+1}y, S^{q+1}x)d(A^{q+1}x, T^{q+1}y) + d(A^{q+1}x, S^{q+1}x)d(A^{q+1}x, T^{q+1}y) \\ + d(B^{q+1}y, T^{q+1}y)d(B^{q+1}y, S^{q+1}x)] \Big\}$$

Now we introduce the following which allows an alternate characterization of the higher-order SC-type contraction

Proposition 4.2.7

Let A, B, S, T be four mappings from a metric space (X, d) into itself, where (A, B) is a higher-order SC-type contraction with respect to (S, T). Put

$$ABST(x,y) := d(Ax, Sx)d(By, Ty) + d(By, Sx)d(Ax, Ty) + d(Ax, Sx)d(Ax, Ty)$$
$$+ d(By, Ty)d(By, Sx)$$

Now for every pair $x \neq y$, define

$$Z := Z(x, y) = \max_{0 \leq v \leq r-1} \beta^{-v} \frac{d(A^v x, B^v y)^2}{ABST(x, y)}$$

then

$$Z = \max_{n \in \mathbb{N} \cup \{0\}} \beta^{-n} \frac{d(A^n x, B^n y)^2}{ABST(x, y)}$$

where $\beta \in [0, \frac{1}{3})$.

Now by the above Proposition, we have the following alternate characterization of the higher-order SC-type contraction

Definition 4.2.8

Let A, B, S, T be four mappings from a metric space (X, d) into itself. We say (A, B) is a higher-order SC-type contraction with respect to (S, T) if the following holds for all $x, y \in X$ and any $r \in \mathbb{N}$

$$d(A^r x, B^r y)^2 \leq Z\beta^r [d(Ax, Sx)d(By, Ty) + d(By, Sx)d(Ax, Ty) + d(Ax, Sx)d(Ax, Ty)$$
$$+ d(By, Ty)d(By, Sx)]$$

where $Z \geq 1$ is given by the previous Proposition and $\beta \in [0, \frac{1}{3})$.

4.3 Main Results

In order to prove the main result we need the following

Lemma 4.3.1

Let A, B, S, T be four self-mappings from a complete metric space (X, d) into itself, satisfying the following conditions

(a) $A^r(X) \subseteq T^r(X)$ and $B^r(X) \subseteq S^r(X)$ for any $r \in \mathbb{N}$

(b) Definition 4.2.8

then the sequence $\{y_n\}$, for any $r \in \mathbb{N}$, defined by

$$y_{2n+1} = T^r x_{2n+1} = A^r x_{2n} \text{ and } y_{2n} = S^r x_{2n} = B^r x_{2n-1}$$

is a Cauchy sequence in X

> **Proof of Lemma 4.3.1**
>
> From the Definition of $\{y_n\}$ and Definition 4.2.8, we have the following
>
> $$\begin{aligned}
d(y_{2n+1}, y_{2n})^2 &= d(A^r x_{2n}, B^r x_{2n-1})^2 \\
&\leq Z\beta^r [d(A^r x_{2n}, S^r x_{2n})d(B^r x_{2n-1}, T^r x_{2n-1}) \\
&\quad + d(B^r x_{2n-1}, S^r x_{2n})d(A^r x_{2n}, T^r x_{2n-1}) \\
&\quad + d(A^r x_{2n}, S^r x_{2n})d(A^r x_{2n}, T^r x_{2n-1}) \\
&\quad + d(B^r x_{2n-1}, T^r x_{2n-1})d(B^r x_{2n-1}, S^r x_{2n})] \\
&= Z\beta^r [d(y_{2n+1}, y_{2n})d(y_{2n}, y_{2n-1}) \\
&\quad + d(y_{2n}, y_{2n})d(y_{2n+1}, y_{2n-1}) + d(y_{2n+1}, y_{2n})d(y_{2n+1}, y_{2n-1}) \\
&\quad + d(y_{2n}, y_{2n-1})d(y_{2n}, y_{2n})] \\
&= Z\beta^r [d(y_{2n+1}, y_{2n})d(y_{2n}, y_{2n-1}) + d(y_{2n+1}, y_{2n})d(y_{2n+1}, y_{2n-1})] \\
&\leq Z\beta^r [d(y_{2n+1}, y_{2n})d(y_{2n}, y_{2n-1}) + d(y_{2n+1}, y_{2n})^2 \\
&\quad + d(y_{2n+1}, y_{2n})d(y_{2n}, y_{2n-1})] \\
&= Z\beta^r [2d(y_{2n+1}, y_{2n})d(y_{2n}, y_{2n-1}) + d(y_{2n+1}, y_{2n})^2] \\
&\leq Z\beta^r [2d(y_{2n}, y_{2n-1})^2 + d(y_{2n+1}, y_{2n})^2]
\end{aligned}$$
>
> From the above we deduce that
>
> $$d(y_{2n+1}, y_{2n})^2 \leq \frac{2Z\beta^r}{1 - Z\beta^r} d(y_{2n}, y_{2n-1})^2$$
>
> or equivalently
>
> $$d(y_{2n+1}, y_{2n}) \leq \sqrt{\frac{2Z\beta^r}{1 - Z\beta^r}} d(y_{2n}, y_{2n-1})$$
>
> Now put $h := \sqrt{\frac{2Z\beta^r}{1 - Z\beta^r}}$, and observe that $h < 1$ since $\frac{2Z\beta^r}{1 - Z\beta^r} < 1$. Now observe for every integer $p > 0$, we have the following
>
> $$\begin{aligned}
d(y_n, y_{n+p}) &\leq d(y_n, y_{n+1}) + d(y_{n+1}, y_{n+2}) + \cdots + d(y_{n+p-1}, y_{n+p}) \\
&\leq h^n d(y_0, y_1) + h^{n+1} d(y_0, y_1) + \cdots + h^{n+p-1} d(y_0, y_1) \\
&\leq (h^n + h^{n+1} + \cdots + h^{n+p-1}) d(y_0, y_1) \\
&\leq h^n (1 + h + h^2 + \cdots + h^{p-1}) d(y_0, y_1) \\
&\leq \frac{h^n}{1 - h} d(y_0, y_1)
\end{aligned}$$
>
> Since $h < 1$, it follows that $\lim_{n \to \infty} h^n = 0$. So taking limits in the above inequality, we deduce that $d(y_n, y_{n+p}) \to 0$. It follows that the sequence $\{y_n\}$ is a Cauchy sequence in X

Now our main result is as follows

> **Theorem 4.3.2**
>
> Let A, B, S, T be four self-maps of a complete metric space (X, d) into itself satisfying the following conditions
>
> (a) $A^r(X) \subseteq T^r(X)$ and $B^r(X) \subseteq S^r(X)$ for any $r \in \mathbb{N}$
>
> (b) Definition 4.2.8
>
> (c) One of A, B, S, T is r-continuous
>
> (d) The pairs (A, S) and (B, T) are r-compatible mappings of type (P)
>
> Then A, B, S, T have a unique common r-fixed point in X

Proof of Theorem 4.3.2

By Lemma 4.3.1, the sequence $\{y_n\}$ is Cauchy, and since X is complete, the definition of $\{y_n\}$ implies $A^r x_{2n} \to z$ and $S^r x_{2n} \to z$ as $n \to \infty$, for some $z \in X$. We assume that A is r-continuous, then $A^r A^r x_{2n} \to A^r z$, and $A^r S^r x_{2n} \to A^r z$ as $n \to \infty$. Since (A, S) is r-compatible of type (P), then $\lim_{n\to\infty} d(A^r A^r x_{2n}, S^r S^r x_{2n}) = 0$, thus we deduce that

$$A^r z = \lim_{n\to\infty} A^r A^r x_{2n} = \lim_{n\to\infty} S^r S^r x_{2n}$$

Now observe we have the following from Definition 4.2.8,

$$\begin{aligned} d(A^r S^r x_{2n}, B^r x_{2n+1})^2 \leq Z\beta^r [&d(A^r S^r x_{2n}, S^r S^r x_{2n})d(B^r x_{2n+1}, T^r x_{2n+1}) \\ &+ d(B^r x_{2n+1}, S^r S^r x_{2n})d(A^r S^r x_{2n}, T^r x_{2n+1}) \\ &+ d(A^r S^r x_{2n}, S^r S^r x_{2n})d(A^r S^r x_{2n}, T^r x_{2n+1}) \\ &+ d(B^r x_{2n+1}, T^r x_{2n+1})d(B^r x_{2n+1}, S^r S^r x_{2n})] \end{aligned}$$

Observe by the completeness of X and the definition of $\{y_n\}$, we have $\lim_{n\to\infty} B^r x_{2n+1} = z$, and $\lim_{n\to\infty} T^r x_{2n+1} = z$, and since $\lim_{n\to\infty} A^r S^r x_{2n} = \lim_{n\to\infty} S^r S^r x_{2n} = A^r z$, we deduce the following, upon taking limits in the above inequality,

$$d(A^r z, z)^2 \leq Z\beta^r d(A^r z, z)^2$$

which is a contradiction, thus, $A^r z = z$. Since $A^r(X) \subseteq T^r(X)$, it follows that there exists $u \in X$ such that $z = A^r z = T^r u$. We show $B^r u = z$. Observe we have the following

$$\begin{aligned} d(A^r x_{2n}, B^r u)^2 \leq Z\beta^r [&d(A^r x_{2n}, S^r x_{2n})d(B^r u, T^r u) \\ &+ d(B^r u, S^r x_{2n})d(A^r x_{2n}, T^r u) \\ &+ d(A^r x_{2n}, S^r x_{2n})d(A^r x_{2n}, T^r u) \\ &+ d(B^r u, T^r u)d(B^r u, S^r x_{2n})] \end{aligned}$$

By the completeness of X and the definition of $\{y_n\}$, we know that $\lim_{n\to\infty} A^r x_{2n} = \lim_{n\to\infty} S^r x_{2n} = z$ and $T^r u = z$, thus taking limits in the above inequality, we deduce that,

$$d(B^r u, z)^2 \leq Z\beta^r d(B^r u, z)^2$$

which is a contradiction. Thus, $B^r u = z$. In particular, $B^r u = T^r u = z$, and since (B, T) is r-compatible of type (P), we deduce that $d(B^r B^r u, T^r T^r u) = 0$, that is, $d(B^r z, T^r z) = 0$ or $B^r z = T^r z$. Since $B^r(X) \subseteq S^r(X)$, there exists $v \in X$ such that $z = B^r z = S^r v$. Now we show $A^r v = z$. Observe we have the following

$$\begin{aligned} d(A^r v, B^r z)^2 \leq Z\beta^r [&d(A^r v, S^r v)d(B^r z, T^r z) \\ &+ d(B^r z, S^r v)d(A^r v, T^r z) \\ &+ d(A^r v, S^r v)d(A^r v, T^r z) \\ &+ d(B^r z, T^r z)d(B^r z, S^r v)] \end{aligned}$$

Since $z = B^r z = S^r v$ and $B^r z = T^r z$, from the above we deduce that

$$d(A^r v, z)^2 \leq Z\beta^r d(A^r v, z)^2$$

which is a contradiction. It follows that $A^r v = z$. In particular, $z = A^r v = S^r v$, and since (A, S) is r-compatible of type (P), we deduce that $d(A^r A^r v, S^r S^r v) = 0$, that is, $d(A^r z, S^r z) = 0$ or $A^r z = S^r z$. Clearly we now have

$$A^r z = B^r z = S^r z = T^r z = z$$

Thus z is a common r-fixed point of A, B, S, T. The uniqueness of the common r-fixed point follows from Definition 4.2.8

4.4 Open Problem

Problem 4.4.1

Inspired by [Y. Rohen, M.R. Singh and L. Shambhu, Common fixed points of compatible mapping of type (C) in Banach Spaces, Proc. of Math. Soc., BHU 20(2004), 77-87] we introduce the following

> **Definition 4.4.1.1**
>
> Let S and T be mappings from a complete metric space X into itself. We say the mappings S and T are r-compatible of type (R) if $\lim_{n\to\infty} d(S^r T^r x_n, T^r S^r x_n) = 0$ and $\lim_{n\to\infty} d(S^r S^r x_n, T^r T^r x_n) = 0$, whenever $\{x_n\}$ is a sequence in X such that $\lim_{n\to\infty} S^r x_n = T^r x_n = t$ for some $t \in X$ and any $r \in \mathbb{N}$

The open problem is to prove the following

> **Theorem 4.4.1.2**
>
> Let A, B, S, T be four self-maps of a complete metric space (X, d) into itself satisfying the following conditions
>
> (a) $A^r(X) \subseteq T^r(X)$ and $B^r(X) \subseteq S^r(X)$ for any $r \in \mathbb{N}$
>
> (b) Definition 4.2.8
>
> (c) One of A, B, S, T is r-continuous
>
> (d) The pairs (A, S) and (B, T) are r-compatible mappings of type (R)
>
> Then A, B, S, T have a unique common r-fixed point in X

Chapter 5

A Higher-Order Fixed Point Theorem under r-Compatible Mappings of Type (K)

5.1 Brief Summary

> **Abstract**
>
> Inspired by higher-order fixed point theory [Clement Ampadu, Fixed Point Theory for Higher-Order Mappings. ISBN: 5800118959925, lulu.com, 2016], we obtain the higher-order version of Theorem 3.1 [Ravi Sriramula and V.Srinivas, A Result on Fixed Point Theorem Using Compatible Mappings of Type (K), Annals of Pure and Applied Mathematics Vol. 13, No. 1, 2017, 41-47]

5.2 Preliminaries

The notion of compatible mappings was introduced by Jungck [G.Jungck, Compatible mappings and fixed points, Intern. J. Math. and Math. Sci., 9(1986) 771-778], and inspired by this concept we introduce the following

> **Definition 5.2.1**
>
> Let S and T be mappings from a complete metric space X into itself. We say the mappings S and T are r-compatible if $\lim_{n\to\infty} d(S^r T^r x_n, T^r S^r x_n) = 0$, whenever $\{x_n\}$ is a sequence in X such that $\lim_{n\to\infty} S^r x_n = T^r x_n = t$ for some $t \in X$ and any $r \in \mathbb{N}$

The concept of weakly compatible mappings have appeared in the literature, and for example see [Ravi Sriramula and V.Srinivas, A Result on Fixed Point Theorem Using Compatible Mappings of Type (K), Annals of Pure and Applied Mathematics Vol. 13, No. 1, 2017, 41-47]. Now we introduce the following

> **Definition 5.2.2**
>
> Two self maps S and T of a metric space (X, d) will be called r-weakly compatible if they r-commute at their r-coincidence point, that is, for any $r \in \mathbb{N}$, if $S^r u = T^r u$ for some $u \in X$, then $S^r T^r u = T^r S^r u$

The concept of compatible mappings of type (A) have appeared in the literature and for example see [Ravi Sriramula and V.Srinivas, A Result on Fixed Point Theorem Using Compatible Mappings of Type (K), Annals of Pure and Applied Mathematics Vol. 13, No. 1, 2017, 41-47]. Now we introduce the following

> **Definition 5.2.3**
>
> Two self maps S and T of a metric space (X,d) will be called r-compatible of type (A), if for any $r \in \mathbb{N}$, $\lim_{n\to\infty} d(S^r T^r x_n, T^r T^r x_n) = 0$ and $\lim_{n\to\infty} d(T^r S^r x_n, S^r S^r x_n) = 0$, whenever $\{x_n\}$ is a sequence in X such that $\lim_{n\to\infty} S^r x_n = T^r x_n = t$ for some $t \in X$

Compatible mappings of type (B) have appeared in the literature, and for example, see [Ravi Sriramula and V.Srinivas, A Result on Fixed Point Theorem Using Compatible Mappings of Type (K), Annals of Pure and Applied Mathematics Vol. 13, No. 1, 2017, 41-47]. Now we introduce the following

> **Definition 5.2.4**
>
> Let S and T be mappings from a metric space X into itself. The mappings S and T will be called r-compatible mappings of type (B) if
>
> $$\lim_{n\to\infty} d(S^r T^r x_n, T^r T^r x_n) \leq \frac{1}{2}[\lim_{n\to\infty} d(S^r T^r x_n, S^r t) + \lim_{n\to\infty} d(S^r t, S^r S^r x_n)]$$
>
> and
>
> $$\lim_{n\to\infty} d(T^r S^r x_n, S^r S^r x_n) \leq \frac{1}{2}[\lim_{n\to\infty} d(T^r S^r x_n, T^r t) + \lim_{n\to\infty} d(T^r t, T^r T^r x_n)]$$
>
> whenever $\{x_n\}$ is a sequence in X such that $\lim_{n\to\infty} S^r x_n = \lim_{n\to\infty} T^r x_n = t$ for some $t \in X$

Compatible mappings of type (P) have appeared in the literature, and for example, see [Ravi Sriramula and V.Srinivas, A Result on Fixed Point Theorem Using Compatible Mappings of Type (K), Annals of Pure and Applied Mathematics Vol. 13, No. 1, 2017, 41-47]. Now we introduce the following

> **Definition 5.2.5**
>
> Let S and T be mappings from a metric space X into itself. The mappings S and T will be called r-compatible mappings of type (P) if $\lim_{n\to\infty} d(S^r S^r x_n, T^r T^r x_n) = 0$ whenever $\{x_n\}$ is a sequence in X such that $\lim_{n\to\infty} S^r x_n = \lim_{n\to\infty} T^r x_n = t$ for some $t \in X$

Compatible mappings of type (K) appeared in [K.Jha, V.Popa and K.B.Manandhar, A Common fixed point theorem for compatible mappings of type (K) in metric space, Intern. J. Math. Sci.and Engg. Appl., 8(I) (2014) 383-391] now we introduce the following

> **Definition 5.2.6**
>
> Two self mappings S and T of a metric space (X,d) will be called r-compatible mappings of type (K), for any $r \in \mathbb{N}$, if $\lim_{n\to\infty} S^r S^r x_n = T^r t$ and $\lim_{n\to\infty} T^r T^r x_n = S^r t$, whenever $\{x_n\}$ is a sequence in X such that $\lim_{n\to\infty} S^r x_n = \lim_{n\to\infty} T^r x_n = t$ for some $t \in X$

5.3 Main Results

Inspired by the mapping contained in Theorem 2.7[Ravi Sriramula and V.Srinivas, A Result on Fixed Point Theorem Using Compatible Mappings of Type (K), Annals of Pure and Applied Mathematics Vol. 13, No. 1, 2017, 41-47] we introduce the following

> **Definition 5.3.1**
>
> Let A, B, S, T be self mappings from a metric space (X,d) into itself. We say the pair (A, B) is an SC-type contraction with respect to (S,T) if the following holds for all $x, y \in X$ and $k \in [0, \frac{1}{3})$
>
> $$d(Ax, By)^2 \leq k[d(Ax, Sx)d(By, Ty) + d(By, Sx)d(Ax, Ty) + d(Ax, Sx)d(Ax, Ty) \\ + d(By, Ty)d(By, Sx)]$$

Definition 5.3.2

Let A, B, S, T be four mappings from a metric space (X, d) into itself. We say (A, B) is a higher-order SC-type contraction with respect to (S, T) if there exists $c_q \in [0, \frac{1}{3})$ such that for all $0 \leq q \leq r - 1$ and $r \in \mathbb{N}$, the following holds for all $x, y \in X$

$$d(A^r x, B^r y)^2 \leq \sum_{q=0}^{r-1} \Big\{ c_q [d(A^{q+1}x, S^{q+1}x) d(B^{q+1}y, T^{q+1}y)$$
$$+ d(B^{q+1}y, S^{q+1}x) d(A^{q+1}x, T^{q+1}y) + d(A^{q+1}x, S^{q+1}x) d(A^{q+1}x, T^{q+1}y)$$
$$+ d(B^{q+1}y, T^{q+1}y) d(B^{q+1}y, S^{q+1}x)] \Big\}$$

Now we introduce the following which allows an alternate characterization of the higher-order SC-type contraction

Proposition 5.3.3

Let A, B, S, T be four mappings from a metric space (X, d) into itself, where (A, B) is a higher-order SC-type contraction with respect to (S, T). Put

$$ABST(x, y) := d(Ax, Sx) d(By, Ty) + d(By, Sx) d(Ax, Ty) + d(Ax, Sx) d(Ax, Ty)$$
$$+ d(By, Ty) d(By, Sx)$$

Now for every pair $x \neq y$, define

$$Z := Z(x, y) = \max_{0 \leq v \leq r-1} \beta^{-v} \frac{d(A^v x, B^v y)^2}{ABST(x, y)}$$

then

$$Z = \max_{n \in \mathbb{N} \cup \{0\}} \beta^{-n} \frac{d(A^n x, B^n y)^2}{ABST(x, y)}$$

where $\beta \in [0, \frac{1}{3})$.

Now by the above Proposition, we have the following alternate characterization of the higher-order SC-type contraction

Definition 5.3.4

Let A, B, S, T be four mappings from a metric space (X, d) into itself. We say (A, B) is a higher-order SC-type contraction with respect to (S, T) if the following holds for all $x, y \in X$ and any $r \in \mathbb{N}$

$$d(A^r x, B^r y)^2 \leq Z \beta^r [d(Ax, Sx) d(By, Ty) + d(By, Sx) d(Ax, Ty) + d(Ax, Sx) d(Ax, Ty)$$
$$+ d(By, Ty) d(By, Sx)]$$

where $Z \geq 1$ is given by the previous Proposition and $\beta \in [0, \frac{1}{3})$

In order to prove the main result we need the following

Lemma 5.3.5

Let A, B, S, T be four self-mappings from a complete metric space (X, d) into itself, satisfying the following conditions

(a) $A^r(X) \subseteq T^r(X)$ and $B^r(X) \subseteq S^r(X)$ for any $r \in \mathbb{N}$

(b) Definition 5.3.4

then the sequence $\{y_n\}$, for any $r \in \mathbb{N}$, defined by

$$y_{2n+1} = T^r x_{2n+1} = A^r x_{2n} \text{ and } y_{2n} = S^r x_{2n} = B^r x_{2n-1}$$

is a Cauchy sequence in X

Proof of Lemma 5.3.5

From the Definition of $\{y_n\}$ and Definition 5.3.4, we have the following

$$\begin{aligned}
d(y_{2n+1}, y_{2n})^2 &= d(A^r x_{2n}, B^r x_{2n-1})^2 \\
&\leq Z\beta^r [d(A^r x_{2n}, S^r x_{2n}) d(B^r x_{2n-1}, T^r x_{2n-1}) \\
&\quad + d(B^r x_{2n-1}, S^r x_{2n}) d(A^r x_{2n}, T^r x_{2n-1}) \\
&\quad + d(A^r x_{2n}, S^r x_{2n}) d(A^r x_{2n}, T^r x_{2n-1}) \\
&\quad + d(B^r x_{2n-1}, T^r x_{2n-1}) d(B^r x_{2n-1}, S^r x_{2n})] \\
&= Z\beta^r [d(y_{2n+1}, y_{2n}) d(y_{2n}, y_{2n-1}) \\
&\quad + d(y_{2n}, y_{2n}) d(y_{2n+1}, y_{2n-1}) + d(y_{2n+1}, y_{2n}) d(y_{2n+1}, y_{2n-1}) \\
&\quad + d(y_{2n}, y_{2n-1}) d(y_{2n}, y_{2n})] \\
&= Z\beta^r [d(y_{2n+1}, y_{2n}) d(y_{2n}, y_{2n-1}) + d(y_{2n+1}, y_{2n}) d(y_{2n+1}, y_{2n-1})] \\
&\leq Z\beta^r [d(y_{2n+1}, y_{2n}) d(y_{2n}, y_{2n-1}) + d(y_{2n+1}, y_{2n})^2 \\
&\quad + d(y_{2n+1}, y_{2n}) d(y_{2n}, y_{2n-1})] \\
&= Z\beta^r [2 d(y_{2n+1}, y_{2n}) d(y_{2n}, y_{2n-1}) + d(y_{2n+1}, y_{2n})^2] \\
&\leq Z\beta^r [2 d(y_{2n}, y_{2n-1})^2 + d(y_{2n+1}, y_{2n})^2]
\end{aligned}$$

From the above we deduce that

$$d(y_{2n+1}, y_{2n})^2 \leq \frac{2Z\beta^r}{1 - Z\beta^r} d(y_{2n}, y_{2n-1})^2$$

or equivalently

$$d(y_{2n+1}, y_{2n}) \leq \sqrt{\frac{2Z\beta^r}{1 - Z\beta^r}} d(y_{2n}, y_{2n-1})$$

Now put $h := \sqrt{\frac{2Z\beta^r}{1 - Z\beta^r}}$, and observe that $h < 1$ since $\frac{2Z\beta^r}{1 - Z\beta^r} < 1$. Now observe for every integer $p > 0$, we have the following

$$\begin{aligned}
d(y_n, y_{n+p}) &\leq d(y_n, y_{n+1}) + d(y_{n+1}, y_{n+2}) + \cdots + d(y_{n+p-1}, y_{n+p}) \\
&\leq h^n d(y_0, y_1) + h^{n+1} d(y_0, y_1) + \cdots + h^{n+p-1} d(y_0, y_1) \\
&\leq (h^n + h^{n+1} + \cdots + h^{n+p-1}) d(y_0, y_1) \\
&\leq h^n (1 + h + h^2 + \cdots + h^{p-1}) d(y_0, y_1) \\
&\leq \frac{h^n}{1 - h} d(y_0, y_1)
\end{aligned}$$

Since $h < 1$, it follows that $\lim_{n \to \infty} h^n = 0$. So taking limits in the above inequality, we deduce that $d(y_n, y_{n+p}) \to 0$. It follows that the sequence $\{y_n\}$ is a Cauchy sequence in X

In order to prove the main result we also need the following

CHAPTER 5. A HIGHER-ORDER FIXED POINT THEOREM UNDER R-COMPATIBLE MAPPINGS OF TYPE (K)

Proposition 5.3.6

Let A and S be r-compatible mappings of type (K) on a metric space (X, d). If one of A or S is r-continuous then we have the following

(a) For any $r \in \mathbb{N}$, $A^r x = S^r x$, where $\lim_{n \to \infty} A^r x_n = \lim_{n \to \infty} S^r x_n = x$, for some $x \in X$

(b) For any $r \in \mathbb{N}$, if there exists $u \in X$ such that $A^r u = S^r u = x$, then $A^r S^r u = S^r A^r u$

Proof of Proposition 5.3.6

Part(a)

Let $\{x_n\}$ be a sequence of X such that

$$\lim_{n \to \infty} A^r x_n = \lim_{n \to \infty} S^r x_n = x$$

for some $x \in X$. Since A and S are r-compatible mappings of type (K), we have

$$\lim_{n \to \infty} A^r A^r x_n = S^r x$$

If A is r-continuous, then

$$\lim_{n \to \infty} A^r A^r x_n = A^r (\lim_{n \to \infty} A^r x_n) = A^r x$$

It follows that $A^r x = S^r x$. Similarly, if S is r-continuous, then we get the same result.

Part(b)

Suppose $A^r u = S^r u = x$ for some $u \in X$, then $A^r S^r u = A^r (S^r u) = A^r x$ and $S^r A^r u = S^r (A^r u) = S^r x$. From (a), we have $A^r x = S^r x$. Hence, it follows that $A^r S^r u = S^r A^r u$

Now our main result is the following

Theorem 5.3.7

Let A, B, S, T be self mappings from a complete metric space (X, d) into itself satisfying the following conditions

(a) For any $r \in \mathbb{N}$, $A^r(X) \subseteq T^r(X)$ and $B^r(X) \subseteq S^r(X)$

(b) Definition 5.3.4

(c) One of the mappings A, B, S, T is r-continuous

(d) The pair (A, S) is r-compatible of type (K)

(e) The pair (B, T) is weakly r-compatible

Then A, B, S, T have a unique common r-fixed point z in X

Proof of Theorem 5.3.7

By Lemma 5.3.5, the sequence $\{y_n\}$ is Cauchy, and since X is complete, there is $z \in X$ such that $\lim_{n \to \infty} y_n = z$. In particular from the definition of $\{y_n\}$ we also have $\lim_{n \to \infty} A^r x_{2n} = S^r x_{2n} = z$ for any $r \in \mathbb{N}$. Suppose A is r-continuous, then $\lim_{n \to \infty} A^r A^r x_{2n} = A^r z$ and $\lim_{n \to \infty} A^r S^r x_{2n} = A^r z$. If A and S are r-compatible of type (K) and one of A, S is r-continuous, then by Proposition 5.3.6, we have $A^r z = S^r z$. Since $A^r(X) \subseteq T^r(X)$, it follows that there exists $u \in X$ such that $A^r z = T^r u$. Now from Definition 5.3.4, we deduce the following

$$d(A^r z, B^r u)^2 \leq Z\beta^r [d(A^r z, S^r z)d(B^r u, T^r u) + d(B^r u, S^r z)d(A^r z, T^r u)$$
$$+ d(A^r z, S^r z)d(A^r z, T^r u) + d(B^r u, T^r u)d(B^r u, S^r z)]$$

Since $A^r z = S^r z$, and $A^r z = T^r u$, it follows that $d(A^r z, T^r u) = 0$ and $d(A^r z, S^r z) = 0$. Thus from the inequality immediately above, we deduce that,

$$d(A^r z, B^r u)^2 \leq Z\beta^r d(A^r z, B^r u)^2$$

or equivalently

$$d(A^r z, B^r u) \leq \sqrt{Z\beta^r} d(A^r z, B^r u)$$

Since $1 - \sqrt{Z\beta^r} > 0$, the above inequality implies $d(A^r z, B^r u) = 0$, that is, $A^r z = B^r u$. It follows that $A^r z = B^r u = T^r u = S^r z$. Now we show $A^r z = z$. Now observe we have the following from Definition 2.4,

$$d(A^r z, B^r x_{2n-1})^2 \leq Z\beta^r [d(A^r z, S^r z)d(B^r x_{2n-1}, T^r x_{2n-1})$$
$$+ d(B^r x_{2n-1}, S^r z)d(A^r z, T^r x_{2n-1})$$
$$+ d(A^r z, S^r z)d(A^r z, T^r x_{2n-1})$$
$$+ d(B^r x_{2n-1}, T^r x_{2n-1})d(B^r x_{2n-1}, S^r z)]$$

We know $\lim_{n \to \infty} B^r x_{2n-1} = z$, $\lim_{n \to \infty} T^r x_{2n-1} = z$, and since $A^r z = S^r z$ implies $d(A^r z, S^r z) = 0$, we deduce from the above inequality that

$$d(A^r z, z)^2 \leq Z\beta^r d(A^r z, z)^2$$

or equivalently

$$d(A^r z, z) \leq \sqrt{Z\beta^r} d(A^r z, z)$$

Since $1 - \sqrt{Z\beta^r} > 0$, the above inequality implies $d(A^r z, z) = 0$, that is, $A^r z = z$. It follows that $A^r z = S^r z = z$ and hence $A^r z = S^r z = T^r u = B^r u = z$. Since B and T are weakly r-compatible, we have $B^r T^r u = T^r B^r u$, and since $T^r u = B^r u = z$, it follows that $B^r z = T^r z$. Now we show $B^r z = z$. Observe from Definition 5.3.4, we have the following

$$d(A^r x_{2n}, B^r z)^2 \leq Z\beta^r [d(A^r x_{2n}, S^r x_{2n})d(B^r z, T^r z)$$
$$+ d(B^r z, S^r x_{2n})d(A^r x_{2n}, T^r z)$$
$$+ d(A^r x_{2n}, S^r x_{2n})d(A^r x_{2n}, T^r z)$$
$$+ d(B^r z, T^r z)d(B^r z, S^r x_{2n})]$$

CHAPTER 5. A HIGHER-ORDER FIXED POINT THEOREM UNDER R-COMPATIBLE MAPPINGS OF TYPE (K)

> **Proof of Theorem 5.3.7 Continued**
>
> We know $\lim_{n\to\infty} A^r x_{2n} = z$, $\lim_{n\to\infty} S^r x_{2n} = z$, and since $B^r z = T^r z$ implies $d(B^r z, T^r z) = 0$, we deduce from the above inequality that
>
> $$d(B^r z, z)^2 \leq Z\beta^r d(B^r z, z)^2$$
>
> or equivalently
>
> $$d(B^r z, z) \leq \sqrt{Z\beta^r} d(B^r z, z)$$
>
> Since $1 - \sqrt{Z\beta^r} > 0$, the above inequality implies $d(B^r z, z) = 0$, that is, $B^r z = z$. It now follows that $T^r z = B^r z = z$, and since $A^r z = S^r z = z$, it follows that z is a common r-fixed point of A, B, S, T. The uniqueness of the common r-fixed point follows from Definition 5.3.4, and the proof is finished

5.4 Open Problems

Problem 5.4.1

Compatible mappings of type (P) have appeared in the literature, and for example, see [V. Srinivas and V. Naga Raju, Common Fixed Point Theorem on Compatible Mappings of Type (P), Gen. Math. Notes, Vol. 21, No. 2, April 2014, pp. 87-94]. Now we introduce the following

> **Definition 5.4.1.1**
>
> Let S and T be mappings from a metric space X into itself. The mappings S and T will be called r-compatible mappings of type (P) if $\lim_{n\to\infty} d(S^r S^r x_n, T^r T^r x_n) = 0$ whenever $\{x_n\}$ is a sequence in X such that $\lim_{n\to\infty} S^r x_n = \lim_{n\to\infty} T^r x_n = t$ for some $t \in X$

The open problem is to prove the following

> **Theorem 5.4.1.2**
>
> Let A, B, S, T be self mappings from a complete metric space (X, d) into itself satisfying the following conditions
>
> (a) For any $r \in \mathbb{N}$, $A^r(X) \subseteq T^r(X)$ and $B^r(X) \subseteq S^r(X)$
>
> (b) Definition 5.3.4
>
> (c) One of the mappings A, B, S, T is r-continuous
>
> (d) The pair (A, S) is r-compatible of type (P)
>
> (e) The pair (B, T) is weakly r-compatible
>
> Then A, B, S, T have a unique common r-fixed point z in X

In order to prove the above, it might be necessary to prove the following and use it in the main result

> **Proposition 5.4.1.3**
>
> Let A and S be r-compatible mappings of type (P) on a metric space (X, d). If both A and S are r-continuous then we have the following
>
> (a) For any $r \in \mathbb{N}$, $A^r x = S^r x$, where $\lim_{n\to\infty} A^r x_n = \lim_{n\to\infty} S^r x_n = x$, for some $x \in X$
>
> (b) For any $r \in \mathbb{N}$, if there exists $u \in X$ such that $A^r u = S^r u = x$, then $A^r S^r u = S^r A^r u$

Bibliography

[1] Clement Ampadu, Fixed Point Theory for Higher-Order Mappings. ISBN: 5800118959925, lulu.com, 2016

[2] M. Bina Devi, Some Common Fixed Point Theorems of Compatible Mappings of Type (A) in Metric Space, Gen. Math. Notes, Vol. 14, No. 1, January 2013, pp. 43-50

[3] G. Jungck, Compatible mappings and common fixed points, Internat. J. Math. Math. Sci., 9(1986), 771-779

[4] G. Jungck, P.P. Murthy and Y.J. Cho, Compatible mappings of type (A) and common fixed points, Math. Japonica, 38(1993), 381-390

[5] Y. Rohen, Th. Indubala, O. Budhichandra and N. Leenthoi, Common fixed point theorems for compatible mappings of type (A), IJMSEA, 6(II) (2012), 323-333

[6] Ravindra K Bisht and Naseer Shahzad, Faintly compatible mappings and common fixed points, Fixed Point Theory and Applications 2013, 2013:156

[7] G. Jungck, "Commuting mappings and fixed points," American Mathematical Monthly, vol. 83, no. 4, pp. 261–263, 1976

[8] Jungck, G: Common fixed points for noncontinuous nonself maps on nonmetric spaces. Far East J. Math. Sci. 4, 199-215 (1996)

[9] Al-Thagafi, MA, Shahzad, N: Generalized I-nonexpansive selfmaps and invariant approximations. Acta Math. Sin. 24, 867-876 (2008)

[10] Pant, V, Pant, RP: Common fixed points of conditionally commuting maps. Fixed Point Theory 1, 113-118 (2010)

[11] Bouhadjera, H, Godet-Thobie, C: Common fixed point theorems for pair of subcompatible maps. arXiv:0906.3159v1 [math.FA] (2009)

[12] Pant, RP, Bisht, RK: Occasionally weakly compatible mappings and fixed points. Bull.Belg. Math. Soc. Simon Stevin 19, 655-661 (2012)

[13] R. Kannan, Some results on fixed points, Bull. Calcutta Math. Soc. 60(1968), 71-76

[14] S. K. Chatterjea, "Fixed-point theorems," Comptes Rendus de l'Academie Bulgare des Sciences , vol. 25, pp. 727–730, 1972

[15] Y. Rohen, M.R. Singh and L. Shambhu, Common fixed points of compatible mapping of type (C) in Banach Spaces, Proc. of Math. Soc., BHU 20(2004), 77-87

[16] M. Koireng Meitei, Leenthoi Ningombam and Yumnam Rohen, Common Fixed Points of Compatible Mappings of Type (R), Gen. Math. Notes, Vol. 10, No. 1, May 2012, pp. 58-62

[17] H.K. Pathak, S.S. Chang and Y.J. Cho., Fixed point theorem for compatible mappings of type (P), Indian J. Math. 36(2) (1994), 151-166

[18] V. Srinivas and V. Naga Raju, Common Fixed Point Theorem on Compatible Mappings of Type (P), Gen. Math. Notes, Vol. 21, No. 2, April 2014, pp. 87-94

[19] Ravi Sriramula and V.Srinivas, A Result on Fixed Point Theorem Using Compatible Mappings of Type (K), Annals of Pure and Applied Mathematics Vol. 13, No. 1, 2017, 41-47

[20] K.Jha, V.Popa and K.B.Manandhar, A Common fixed point theorem for compatible mappings of type (K) in metric space, Intern.J. Math. Sci.and Engg. Appl., 8(I) (2014) 383-391

www.ingramcontent.com/pod-product-compliance
Lightning Source LLC
Chambersburg PA
CBHW051103180526
45172CB00002B/755